SIP DEMYSTIFIED

SIP Demystified

Gonzalo Camarillo

McGraw-Hill
New York Chicago San Francisco
Lisbon London Madrid Mexico City
Milan New Delhi San Juan Seoul
Singapore Sydney Toronto

McGraw-Hill
*A Division of The **McGraw-Hill** Companies*

7 8 9 0 QSR/QSR 0 9 8 7 6 5

ISBN 0-07-137340-3

The sponsoring editor for this book was Marjorie Spencer and the production supervisor was Pamela A. Pelton. It was set in Century Schoolbook by MacAllister Publishing Services, LLC.

Printed and bound by *R.R. Donnelley & Sons.*

This book is printed on recycled, acid-free paper containing a minimum of 50 percent recycled de-inked fiber.

To my family, for their continuous support and encouragement. The education they have given me is the best heritage one could ever get.

ACKNOWLEDGMENTS

This book was possible thanks to my management in Ericsson Finland, namely Christian Engblom, Jussi Haapakangas, Rolf Svanbäck, Roger Förström, and Stefan Von Schantz. Their encouragement was essential in the first stages of the book. They together with Carl Gunnar Perntz and Olle Viktorsson, from Ericsson Sweden, let me go to Columbia University in New York to work together with Professor Schulzrinne when I was finishing writing SIP Demystified.

Professor Schulzrinne's advice and guidance has been very important to me since I began working on SIP-related issues, when SIP was still a brief Internet draft within the MMUSIC working group.

Miguel Angel García and Jonathan Rosenberg provided guidance and detailed comments. Their reviews of the manuscript contributed to improve the final product.

Last but not least, Marjorie Spencer, my editor at McGraw-Hill, did a terrific work on the manuscript. She deserves recognition for her impact on the final manuscript. She provided new ideas and different points of view that helped make the technical explanations contained in this book clearer to readers with all types of backgrounds.

CONTENTS

PREFACE

The *Session Initiation Protocol* (SIP) has gained a lot of attention over the last few years. Lately, the decision to use SIP as the signalling protocol to provide IP multimedia services in the third generation of mobile systems has dramatically increased the number of people interested in knowing about SIP. SIP is the protocol that will merge together the cellular and the Internet worlds. It will provide ubiquitous access to all the services that have made the Internet so successful . Users will be able to combine traditional Internet services such as e-mail and the Web with newer services such as multimedia and instant messaging.

Although the services that SIP can provide are relatively well known, there is a lack of knowledge about the protocol itself. SIP is seen by many people as a protocol that can resolve every problem one could imagine, when in reality, SIP has a limited well-defined scope. During all the years that I have been working on SIP standardization, I have heard this and many other misconceptions several times. That was the main reason that pushed me to write this book. This book intends to clarify the philosophy behind SIP.

In order to have a good understanding of a protocol such as SIP, it is necessary to be able to answer three simple questions: *what*, *how*, and *why*. This book deals with all three of them, but sets a special focus on the last one: why. The reason for doing so is from my own experience of speaking with several engineers and programmers. I was surprised to meet people that have a great knowledge about what SIP does and how it does it, but who did not understand the philosophy behind the protocol. They did not know why SIP was designed like it was. They understood the protocol details but were missing the overall picture. The trees did not let them see the forest. The why part is also very useful for business managers who do not need to know the protocol details in depth, but need to understand why to use a particular technology in their products. It would be sad if SIP was used just because it is fashionable and not for all its good features.

In order to understand why SIP is a good signaling protocol, it is necessary to know the paradigm behind it and compare it with another paradigm to see its advantages. That is why Chapter 1, "Signaling in the Circuit Switched Network," provides an introduction to traditional telephone signaling. This brief introduction helps the reader understand why a paradigm shift was needed, and the advantages and disadvantages of the Internet paradigm.

An introduction to packet switching and to IP has been added at the beginning of Chapter 2, "Signalling in the Packet-Switched Network." This introduction is intended for those professionals who have experience in the telecom world and are trying to jump into the new datacom technologies. They will find advantages and disadvantages of packet switching networks and why IP, and not other network layer protocols, is used to implement packet based services in modern networks.

The remainder of Chapter 2 and Chapter 3, "The Internet Multimedia Conferencing Architecture and How Protocols Mature," set SIP in its context. They describe how SIP interacts with other protocols (the Internet multimedia conferencing architecture), and how SIP standardization is carried out within the *Internet Engineering Task Force* (IETF). This gives the reader an idea of the different maturity levels of the different SIP extensions and what they mean. Knowing the Internet multimedia conferencing architecture is useful in order to understand the scope of SIP and of the rest of the protocols that belong to the architecture. All these protocols interact among them to provide multimedia services to the users.

Chapter 4, "The Session Initiation Protocol," through Chapter 6, "Extending SIP: The SIP Toolkit," deal more with what and how, without forgetting about why. However, these two concepts are kept separate as much as possible. It is important to distinguish between the functionality provided by SIP, and the protocol details of how this functionality is achieved. Understanding first what the protocol does, it is easier to study how it is implemented. Chapter 4 deals with what SIP does wheras Chapter 5, "SIP: Protocol Operation," explains the protocol syntax. This distinction is also present in Chapter 6, where several SIP extensions are explained. Every extension is clearly divided into two sections: the first one explains what the extension does, and the second deals with its implementation.

Finally, Chapter 7, "Building Applications with the SIP Tookit," provides examples of architectures that have chosen SIP as a signaling protocol, such as 3G or PacketCable.

After reading this book you will have a good understanding of the three aspects of SIP: what, how, and why. You will be able to understand its role in different architectures and its interactions with other protocols. Furthermore, you will be able to decide if SIP is the proper tool to use in order to resolve your problem and if so, what other tools you will also need to build the architecture that suits your application better. It is good to always keep in mind that rather than being in an isolated protocol, SIP is part of the Internet multimedia conferencing architectures—a set of protocols that combined can be used to provide multimedia services.

FOREWORD

Unbeknownst to those outside the field of telecommunications, a quiet revolution is taking place. This revolution is aimed at overthrowing the decades-old technologies, now long past their prime, which are the cornerstone of today's wired and wireless telecommunications networks. This revolution will free people from the high cost and low value-add of many telecommunications services, and bring them into the low cost and high value-added services that are the norm on the Internet. The revolution is not being fought with swords or guns, but rather, with technologies—Internet technologies—which are being used to completely redefine the architecture of telecommunications networks. At the lead of this quiet revolution is the *Session Initiation Protocol* (SIP), an Internet standard developed by the *Internet Engineering Task Force* (IETF).

Ever heard of SIP? Probably not—and that's the problem. Up until now, knowledge of SIP and its related technologies has been the domain of the technology elite. However, the changes in the telecommunications industry that SIP is causing are important for many people—from technology managers to businessmen to enterprise network administrators. These people don't need to know the details of the technology, but they need to appreciate its importance and understand how it might impact their work. That is where this book fits in. *SIP Demystified* is not a book for software developers or protocol engineers. It's a book for a much broader audience that helps provide context for SIP and an appreciation for its purpose, basic operation, and relationship to other protocols and technologies.

Gonzalo Camarillo does an excellent job walking the fine line between technical completeness and technology overviews. Gonzalo has been an important contributor to the SIP revolution. He is the author of several of the key documents being developed within the standards bodies. He is an active contributor on the mailing lists, and a good teacher for those who have basic questions. It is this combination of technical depth and teaching skills that have resulted in a fine book, which I wholeheartedly recommend to anyone who asks the question, "what is SIP, and why is it important?"

As one of the co-authors of SIP, I have dedicated several years of my life to its development. I have seen it grow from an academic technology to a force that is changing the way telecommunications will work in the years to come. It is truly impressive that throughout its growth, the fundamental goals and principles on which the technology was built have not changed. This has happened because of the cadre of people who have come to believe in the vision SIP presents, and who have worked to promote it within their organizations. Gonzalo has captured that vision in a book, and so I encourage you, reader, to turn the page and learn more about what the future of telecommunications will look like.

DR. JONATHAN ROSENBERG
CO-AUTHOR, *SESSION INITIATION PROTOCOL*
CHIEF SCIENTIST, DYNAMICSOFT INC.

CHAPTER 1

Signalling in the Circuit-Switched Network

The telephone network, also known as the *Public Switched Telephone Network* (PSTN), reaches almost every country in the world. We can find telephone sets almost in every house, including simple analog phones, more advanced *Integrated Services Digital Network* (ISDN) phones, cordless phones, cellular phones, and even satellite phones. Among technologies, telephone systems are astonishingly widespread.

All these phones have something in common. They use a circuit-switched network to communicate between them. The PSTN has been around for a long time, and its users are happy with it. People can talk to each other across oceans and still feel in proximity. Even in long-distance calls between two mobile phones, it is possible to understand what the other person is saying relatively well.

Besides, the PSTN is a highly-reliable network. When somebody picks up a phone in order to make a call, most of the time the call goes through. Telephone switches rarely crash, and when they do, backup systems take over to continue providing service to the users.

In short, people trust the PSTN. They have grown to expect trouble-free everyday use, and they are confident that in the case of an emergency, the PSTN will connect them to an ambulance, the police, or the fire department.

Taking into account all these features of the PSTN, one might think that the procedures and mechanisms used by the telephone network would be emulated in any network providing a similar service. Since the PSTN works so well, let us model any voice network after it. Specifically, let us use PSTN-like signalling protocols—they work well.

This assumption is wrong, however, and we will see why for a different environment, such as the Internet, PSTN-like protocols are not suitable. Material differences in the PSTN architecture and the Internet architecture require completely new signalling protocols rather than the evolution of trusty, old ones.

This chapter contains a brief PSTN-signalling history that explains how signalling protocols have evolved from analog to digital. We will see that beyond the differences in transport technology used by the PSTN and the Internet—the former is circuit-switched while the latter is packet-switched—differences in their respective paradigms for governance and operation make different signalling design mandatory.

The Origins of Circuit-Switching

A telephone network aims to provide its users with certain services. Many services fall into the category of telephone services, but the first and most important one is the transmission of voice between users. Any user of the network has to be able to call any other user attached to it. All telephone networks must fulfill this requirement.

The first telephone network that enabled two users to communicate consisted of two telephone sets joined by a cable. That system provided both users of the network with adequate voice service, but as soon as more users wanted to be able to use the network, the number of telephone sets increased. The best way to add a new user to the network was to install new cables from the telephone set of the newcomer to all the already-installed telephone sets, creating what is known as a fully meshed topology, where every telephone set is directly connected to every other telephone (see Figure 1-1).

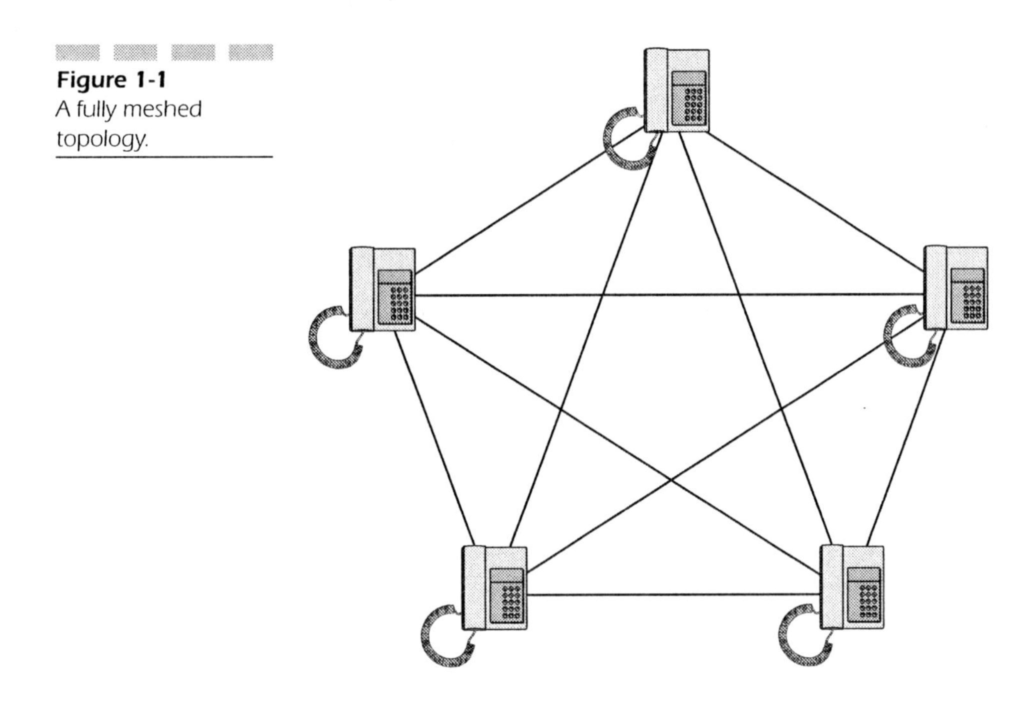

Figure 1-1
A fully meshed topology.

Unfortunately, a fully meshed topology presents a number of disadvantages in the telephone context. Failure to scale is a very important drawback. An individual connection from each user to any other user who potentially can be called has to be established. A physical connection to every single user is affordable as long as the system is small, but as the number of users grows, the number of connections grows in an exponential manner. Cables are expensive and difficult to install. Therefore, adding new users to the network becomes very expensive very quickly.

Cost aside, if a single telephone set has as many cables connected to it as there are users in the network, routing becomes complicated to manage. Each telephone set has to maintain a huge routing table that indicates which cable corresponds to which destination in order to route the calls.

Due to their limitations of scalability and management, early systems evolved from a fully meshed topology to a star topology. In a star topology, all telephone sets are connected to a single central unit called a switch (see Figure 1-2) and all calls are routed through it from source to destination. The switch connected cable from the call's originator to cable at the call's receiver. At first it was operated manually, by a now-quaint figure known as a "switchboard operator" who plugged wires into sockets.

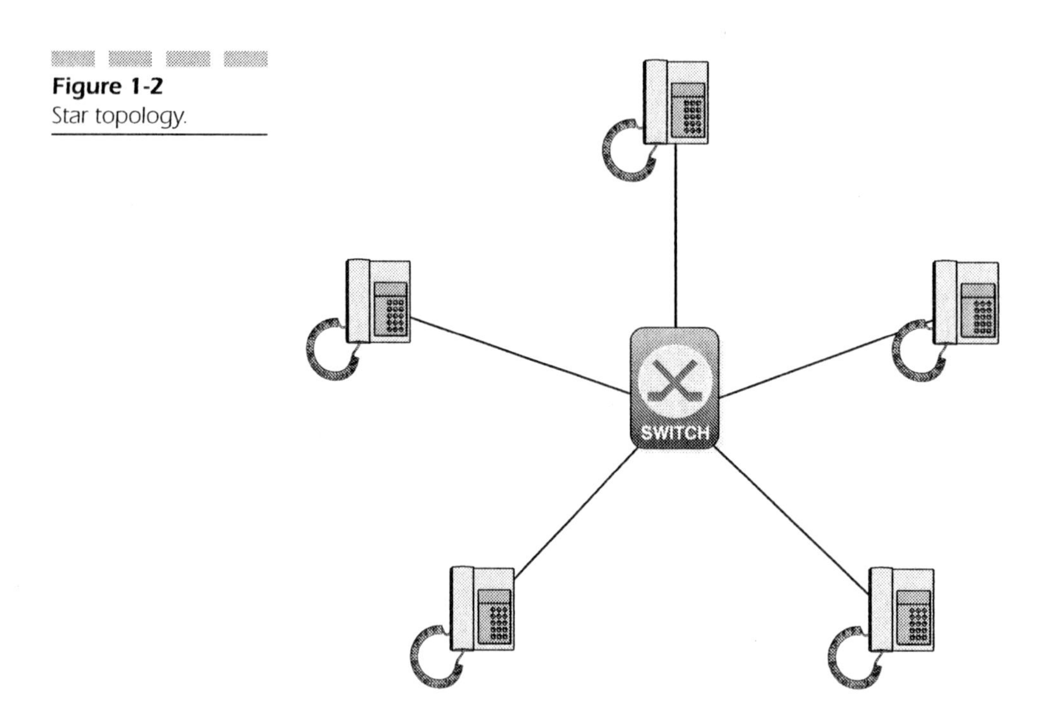

Figure 1-2
Star topology.

Old though it may be, this technology is still in use and still referred to as circuit switching. Obviously, circuit switching facilitates the addition of new users to the system, because it reduces the "cost" to a single cable between the switch and the new telephone set. Management of the system is streamlined. The switch still has to handle a number of circuits, but the user equipment—the telephone set—remains simple.

The next step in circuit-switching is to connect together several switches building a circuit-switched network (see Figure 1-3). In such a network, each user's telephone set is connected to the switch nearest to it. That switch is connected to another, and so on across the system.

In telephony, switches are also referred to as exchanges. Therefore, the closest switch to a telephone set is called its local exchange. Calls between two telephones are routed first to the local exchange of the caller and then traverse other exchanges until reaching the local exchange of the callee, which can announce the call to him by ringing his telephone. Circuit-switching overcame the limitations of the first telephone system and remains the dominant technology for voice transmissions even now. But circuit-switching technology evolved repeatedly from that very first network, and at present, several hierarchical levels of switches are implemented all together to make up the PSTN, a worldwide communication network.

Figure 1-3
Hierarchical topology.

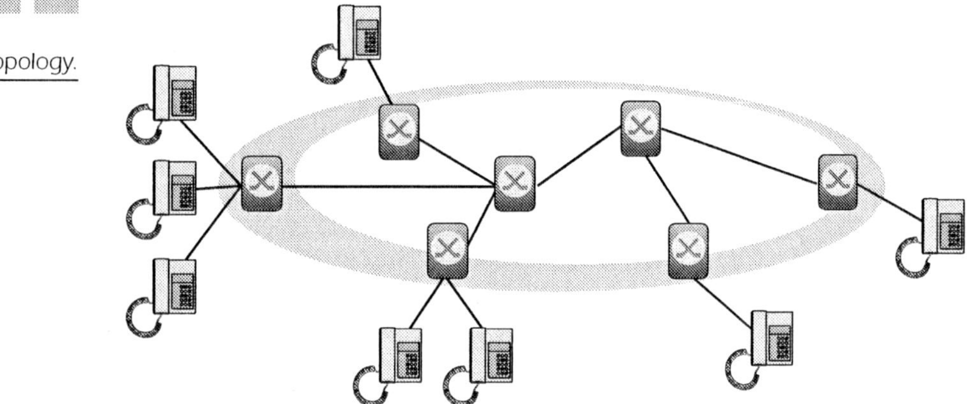

Characteristics of Circuit-Switching

Besides voice transport, circuit-switching is regularly used to transport different types of traffic. Circuit-switched networks transport data between computers and control signals between terminals, for instance. However, no matter which kind of traffic is transported, the user equipment is called a "terminal" and the set of switches is called the "network."

Usually terminals are manipulated by the end user while the network is managed by the service provider. The network establishes the communication path between terminals, whose role is to receive information from the user and transmit it to the network in the proper format. Terminals also perform the opposite task: receiving information from the network and transmitting it to the user. In both cases, it's the terminal alone that examines the content of the information being transmitted and performs the necessary operations on it. Conversely, switches establish a dedicated path between two or more terminals and are not concerned with what they transmit between them.

We've seen that this dedicated path can be as rudimentary as a set of cables strung between terminals and switches. In more advanced systems, paths consist of, for instance, a frequency inside a cable, a time slot, or a wavelength inside an optical fiber. No matter how sophisticated the path, however, the key to circuit-switching remains the same: switches are oblivious to the contents of the messages they are transporting. They maneuver connections based solely on the position of the cables (that is, cable #1 has to be connected to cable #4), the frequency (that is, output from cable #1 at 400 Hz has to be transmitted on cable #4 at 600 Hz), and the time slots inside a frame. We will see in Chapter 2, "Packet Switching, IP, and the IETF," that this is the main difference between circuit-switching and packet-switching.

Strengths of Circuit-Switching

Circuit-switching provides some good features. These systems are very fast. Since switches do not examine the contents of the transmission, the decision on where to send the information received at a certain interface is made just once, at the beginning of connection, and remains the same for the duration. Thus, the delay introduced by a switch is almost negligible.

The dedicated transmission path between terminals provided by circuit-switching networks is well suited for the analog transmission of voice. After a circuit has been established, the transmission delay is small and is kept constant through the duration of the connection. Its suitability for analog transmissions is of course the key reason why circuit-switching technology spread so far and wide in pre-digital days. We will see that when new traffic patterns were introduced, including the transmission of data between computers, circuit-switching began showing its limitations.

Weaknesses of Circuit-Switching

Before an information exchange between terminals can take place, a path through the network must be established. Setting up a path takes time, and the actual transmission is delayed until the path establishment phase is finished (see Figure 1-4).

Note that path establishment delay is distinct from switch delay. By its nature, establishment delay occurs before any transmission takes place. The delay introduced by the switches during the transmission of user data occurs after the path has been established.

Once a dedicated path is established between two terminals, the resources associated with it cannot be used for another connection until the path is torn down. Therefore, even if at some point both terminals stop transmitting, the path stays open and resources allocated to the connection through all the switches along the path remain in use. That's a non-efficient use of available resources, to say the least. This limitation, while not a big problem in analog voice transmission, becomes very important in digital transmissions and particularly in data transmissions between computers.

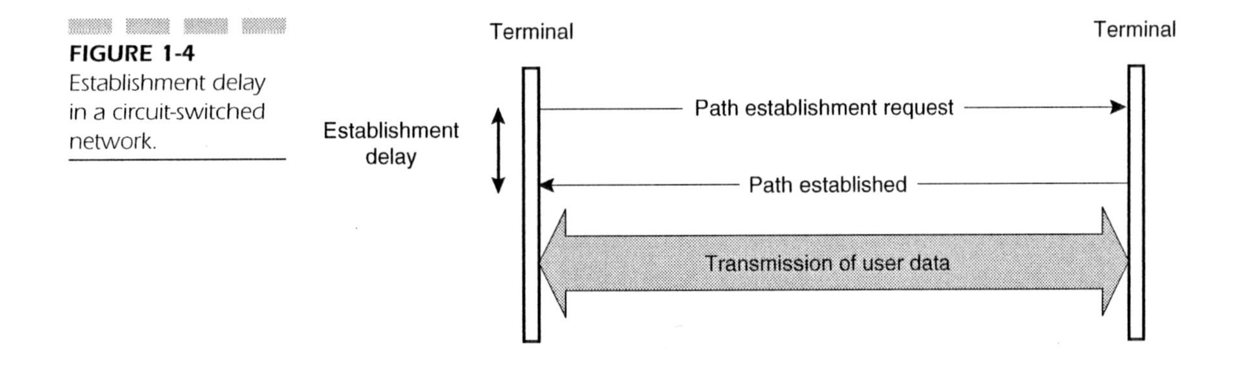

FIGURE 1-4 Establishment delay in a circuit-switched network.

Two terminals in the act of exchanging data are actually idle most of the time, and so are the resources allocated in the switches. Wherever traffic exchange is bursty and non-uniform, circuit-switching does expend resources inefficiently. We will see that packet-switching overcomes this limitation (but only at the cost of slower switching).

To connect terminals across a network, the switches need to know when to establish the path, when to tear it down, what resources it will need to command (that is, number of circuits to be used), and where to route the call. All this information is exchanged between the switches within the network, and also communicated between the terminals and the network, on every call. Collectively, the information related to the control of the connection is referred to as signalling.

Introduction to Signalling

The purpose of signalling is to exchange control information between systems of any kind. Familiar applications of signalling include, among others, the control of railway traffic and air traffic. When a pilot asks for permission to land at an airport, he is exchanging signalling information with the control tower. The control tower then provides the airplane with a time slot and a runway on which to land. The presence of signalling in telephony implies that two kinds of traffic are exchanged in a telephone call: signalling traffic—controlling the establishment and release of the voice path —and voice traffic. We commonly refer to these different kinds of traffic as planes. Thus, two planes can be found in a telephone call: the control plane and the user plane. The control plane handles the procedures controlling the user plane and the user plane handles the actual data or voice transmission (see Figure 1-5).

The evolution of the signalling plane has been tightly tied to the evolution of the user plane. New features in the switches could not be exploited without a signalling system capable of taking advantage of them. New signalling systems are designed to make optimal use of the latest advances in the user plane. However, there have also been cases where new signalling systems were introduced without any progress in the user plane. In these situations, some other gain—simplicity, efficiency, robustness—was at stake. To understand how signalling protocols have evolved from the first

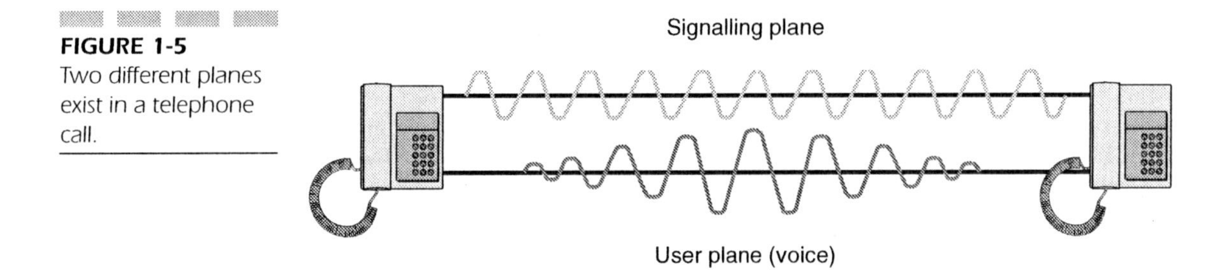

FIGURE 1-5 Two different planes exist in a telephone call.

telephone system back in the nineteenth century up to the present is to understand which gains were achieved and what exactly triggered protocol design at a certain point of time. SIP is both one such protocol and also a departure from the evolutionary path.

Local and central battery systems After years of using telegraphs, on March 10, 1876, Alexander Graham Bell patented electrical voice transmission using a continuous current, and the telephone was born.

Signalling in the early telephone systems was very simple. When the user picked up the receiver, a circuit closed. The terminal then supplied a current to the circuit. The action of closing the circuit meant its seizure. A human operator at the switchboard answered upon seizure of the circuit and routing was accomplished by telling the operator the identity of the callee. The operator switched the call manually. This is the Local Battery system.

Local Battery systems were problematic. Terminals contained batteries, making maintenance more difficult because it was left in the hands of users. Since battery technology was not as advanced as in modern terminals, users found the job complicated and were not eager to take it on. The Central Battery system was devised to make things easier on the users. Central Battery systems supply current to the terminals from the exchange. System automation was the next step in making telephones easier to use. From the installation of the very first exchange in 1878, circuit-switching had been performed manually. In 1889, Almon B. Strowger applied for a patent for an automatic telephone electromechanical switch, and human intervention was no longer needed within the exchanges. Users acquired a dial tone upon seizure of the circuit. After the provision of a dial tone by the local exchange, terminals forwarded the digits of the callee's telephone number. With this information in hand, the local exchange routed the call towards the proper destination.

In these first automated systems, the only signalling direction was forward, for example, from the caller's terminal to the switch. Any information that had to be transmitted backwards to the user was sent through the user plane. So if the callee was busy, for instance, a busy tone was sent through the user plane, and it was up to the user to react by hanging up. The control plane never knew if the user hung up because he or she decided to abort the call before the callee answered or because the callee was busy, which proved to be a disadvantage down the line.

DC and AC Analog Systems The next signalling systems to be developed were the DC and AC analog systems, so-named according to the type of current they used. The gain was that DC and AC systems allowed signalling backwards. Having signalling in the callee-to-caller direction represents an important saving in some situations, as Figure 1-6 illustrates.

The caller dials a phone number, and each switch in the path between caller and callee reserves a circuit to transport the voice content of this call. When the signal arrives at the callee's local exchange, the final switch in the path, it locates the connection to the callee's terminal. The switch checks the availability of this connection and notices that it is being used for another call at that moment. In a system that does not enable signalling backwards, the local exchange then generates a busy tone. This tone is transmitted to the caller so he or she knows to give up on the call. The caller is expected to hang up after hearing this tone but may not immediately do so. Once he does, signalling indicating the caller has gone is sent forward towards the callee's local exchange, which at last triggers the release of all the circuits along the path.

It works, but it entails that many circuits in the network remain engaged by this connection even after the exchange reports that the caller is unavailable. DC and AC analogue systems make better use of network resources. Sending busy status through the control plane allows the user plane, circuit-switched path for the call to be released. The busy signal that the caller hears is instead generated by his or her local exchange, and no resources are reserved unnecessarily while the user is informed about the status of the call and prepares to react (see Figure 1-7).

Thus, having signalling in both directions helps manage the network more efficiently. The control plane receives more, and more accurate, information about the status of the call.

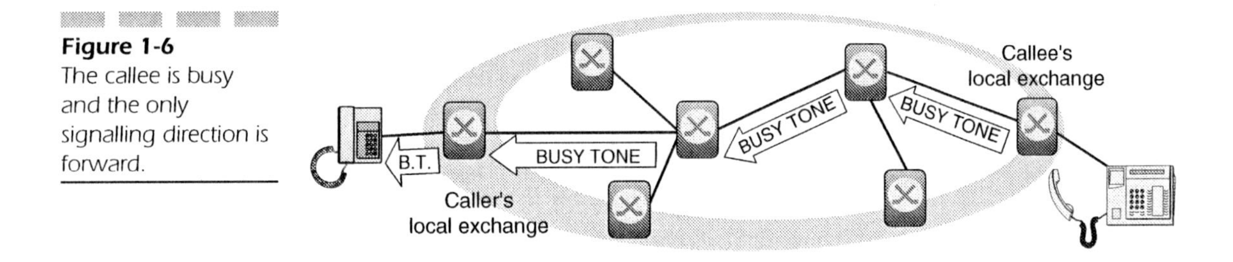

Figure 1-6 The callee is busy and the only signalling direction is forward.

FDM and In-band Signalling

Circuit-switching continued to develop in the next decades. Where each circuit used for telephony consisted of a copper wire, so a switch handling 500 circuits needed 500 cables. *Frequency Division Multiplexing* (FDM) made it possible to use a single wire for several simultaneous calls. In FDM systems, every voice path occupies a different bit of the frequency spectrum (see Figure 1-8), in effect redefining a circuit as a frequency within a physical cable instead of the cable itself. FDM was developed around 1910, but wasn't actually implemented until 1950. At that point a coaxial cable was capable of providing 1,000 circuits—a solid gain.

FDM systems demanded new signalling procedures; the signals within a cable between two switches were needed to control many circuits instead of one. Switches now needed a way to relate the control plane of a specific call to its user plane. The solution would prove to be in-band signalling systems, meaning that signalling is carried on the same frequency as speech—they travel together in the same circuit.

In simplest terms, seizure of the circuit is marked by a pulse that launches signal along the very path that later will transport voice. In-band

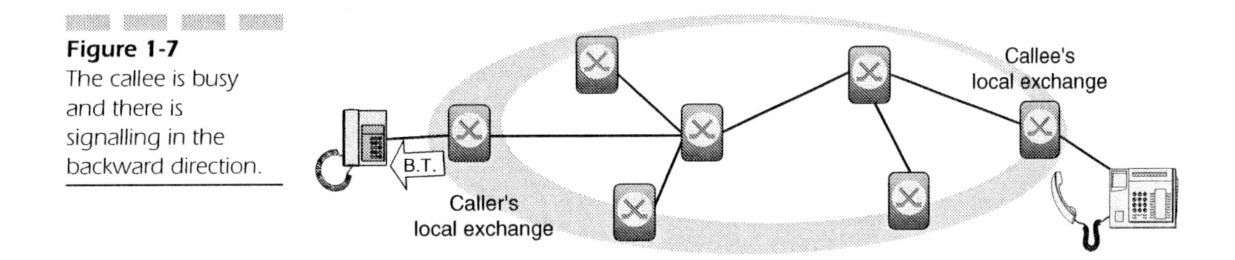

Figure 1-7 The callee is busy and there is signalling in the backward direction.

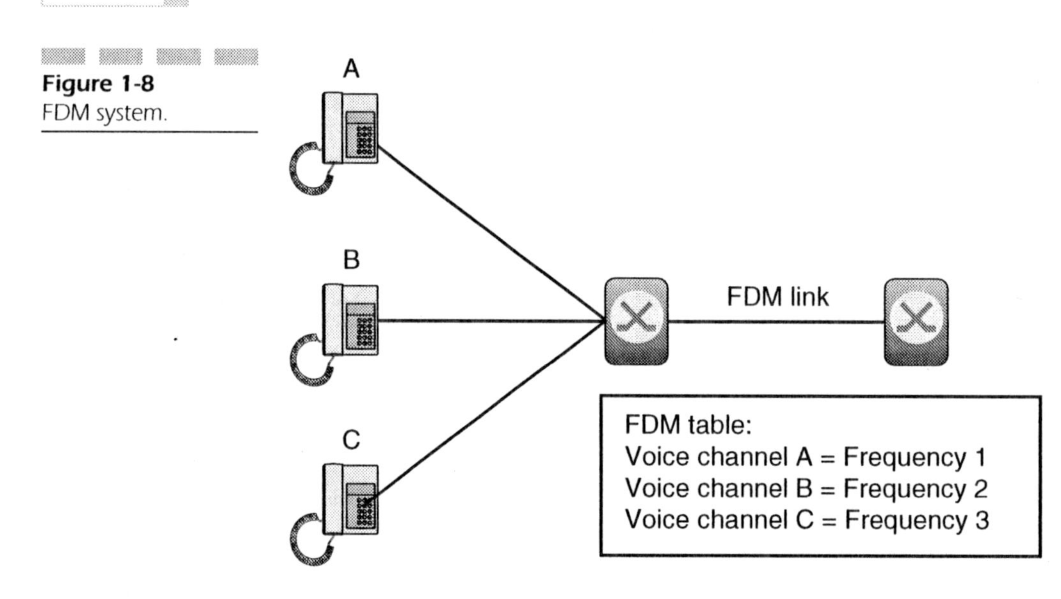

Figure 1-8
FDM system.

signalling is still used at present in some countries, mainly for international connections.

Although FDM reduced the cost of telephone systems by putting many connections on the same cable, it proved to be an expensive technology. FDM uses analog filters to keep different channels separate. The filters are costly to maintain. They have to be checked periodically—as often as once a month—to keep them working in the required frequency.

Analog Transmission

From the first, voice transmission was always analog. Another way to say this is that transmission consisted of sending continuous signals, such as voice, without regard to their contents (see Figure 1-9). The signal to be transmitted is represented by an electromagnetic wave that moves through the cable, bearing the information. A recipient needs to recover the original signal from the electromagnetic wave received in order to possess the message.

Continuous signals have different frequency components, but every signal concentrates most of its energy within a certain range of frequencies—the so-called spectrum of the signal. Frequency components outside the spectrum contain little energy, and thus contribute less to the interpretation of a certain signal. As a result, it is usually necessary to transmit only

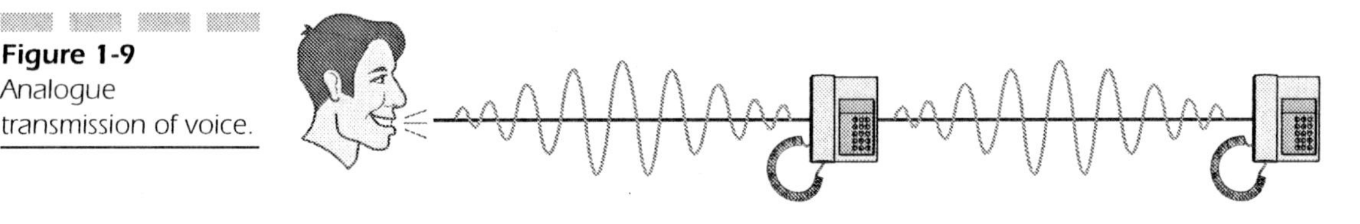

Figure 1-9 Analogue transmission of voice.

the frequency components inside the spectrum for recovery of the original signal, or, to be precise, a signal very similar to the original.

In analog transmission, the electromagnetic wave obtained from a signal contains all frequency components of the spectrum. The standard spectrum for voice transmissions ranges from 300 Hz to 3,400 Hz, and voice transmitted using this range is perfectly understandable by the recipient.

The main advantage of analog transmission is the absence of delay. Because switches in the transmission path do not add delays to the signal, transmission speed is effectively the speed of the electromagnetic wave in the medium. This feature suits interactive communications—where the parties involved are active senders and receivers—exceptionally well.

Digital Transmission

By now you've probably noted a pattern developing whereby advances in the user plane prompt advances in the signalling plane to take advantage of them, in turn re-enabling the user plane for growth. The next step in telephony evolution was accordingly the introduction of digital transmission. Digital transmission changed the way voice was transported and, in so doing, made signalling protocols evolve dramatically.

Just how did this work? We explained earlier that analog signals take on continuous values. Conversely, digital signals take on discrete values. A good example of a digital signal is text. The text is made of letters, and since every letter is chosen from a set of letters—the alphabet—so, the discrete values that any letter can take range from "a" to "z."

Just two values are used in voice digital transmissions: 0 and 1. Therefore, the information exchanged between systems consists of a stream of 0s and 1s. Several ways exist for encoding this kind of stream on the medium being used. One way, among many, is to assign a constant voltage level to 1 and a different voltage level to 0. The receiver can recover the original digital stream by analyzing the voltage levels received.

However, voice is not a digital signal. It is an analog acoustic signal that takes on continuous values in the time domain and requires translation before it can be transported. In order to transport an analog signal over a digital link, a coder-decoder is needed. A coder receives an analog signal as input and produces a digital signal as output. This process can be performed using different algorithms. The algorithms for converting voice into a digital stream are referred to as audio codecs (see Figure 1-10). The most widely used is the G.711 codec, also known as *Pulse Code Modulation* (PCM), which produces 64-kbps streams.

Strengths of Digital Transmission Digital transmission presents many advantages over analog transmission. First, digital equipment is cheaper. Stream manipulations can be performed by relatively cheap computers, while analog switches need expensive electromechanical gear. Since the first digital exchange was installed in 1960 in the U.S., the price of computers has dropped drastically. Second, digital transmission quality is higher. Analog transmission suffers from signal attenuation, that is, the strength of the signal decreases with distance. Exchanges employ amplifiers to keep signal strength above an acceptable level, but to do so, they have to amplify the whole signal. (Remember that exchanges are unable to examine the contents of analog transmissions). Amplifying the signal as a whole has the unfortunate result of also amplifying noise, producing a distortion in the signal received at the destination.

By contrast, digital systems use repeaters rather than amplifiers. A repeater decodes the stream received and reproduces it on the outgoing side. Thus the outgoing signal is completely new, and its strength is independent of the strength of the signal in the previous link. No noise is added to the signal in a digital exchange by the transmission process.

Weaknesses of Digital Transmission Regarding delay, digital systems perform worse than analog systems. The processing of digital signals is

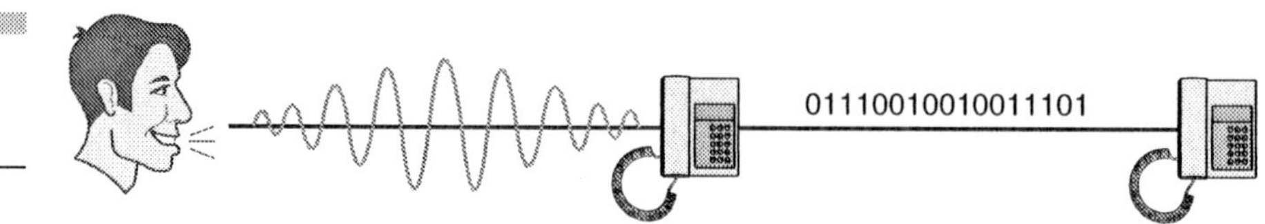

Figure 1-10 Digital transmission of voice.

always slower than the transmission of an electromagnetic wave by an analogue exchange. We will see that the multiplexing mechanism used in digital transmissions (TDM) also introduces some delays, since the data has to wait in a buffer for the time slot assigned to.

Although buffer delays exist in every digital network, modern switches have very high processing power and perform all these operations very rapidly. Therefore, delays introduced by digital transmission are negligible for most applications. For instance, for a voice application, transmission delays higher than 300 ms round-trip make it difficult to maintain a normal conversation. The delay introduced by digital switching is far below this threshold.

A more problematic disadvantage of digital systems for transmitting voice signal is the complexity of the terminal, which has to include a coder-decoder to convert the acoustic signal into a digital stream. This mechanism makes digital terminals trickier and more expensive than analog ones. To avoid terminal complexity, many digital networks implement an analog interface towards the terminal. The interface between the terminal and the local exchange, which is called a local loop or subscriber line (see Figure 1-11), enables the user to retain his or her trusty analog terminal and still connect to a digital local exchange. the coding and decoding of analog signals takes place within the local exchanges of the parties conversing.

Reading between the lines, it can be seen that at the time systems were migrating from analog to digital, the philosophy of the PSTN already was to design simple user equipment and complex networks. There were some additional reasons for that back then, including backwards-compatibility and the price of the terminals, but undeniably this design style persists into the present.

Time Division Multiplexing

FDM, besides being a expensive technology, also limits multiplexing capacity. Different channels have to use frequencies sufficiently separated to avoid interference, and the usable frequency range is defined by the bandwidth of the medium and by the attenuation. (Attenuation is a function of the frequency: the higher the frequency, the more severe the attenuation.) Because high attenuation requires high amplification in the exchanges, the noise that accompanies the analog signal is also highly amplified, degrading quality.

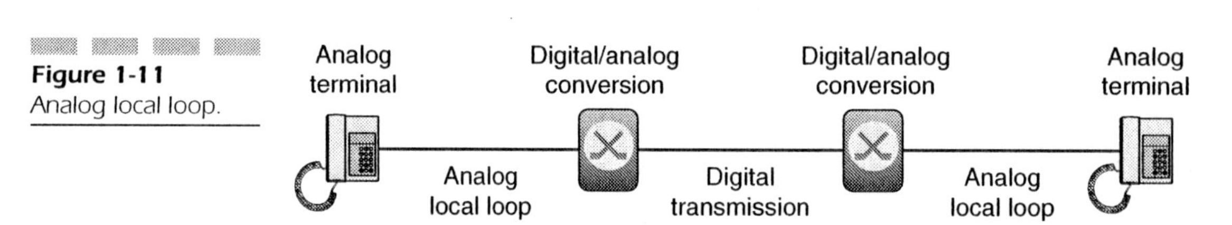

Figure 1-11
Analog local loop.

Multiplexing limitations are a real drawback for the network and helped to spur migration away from FDM. The advent of new high-bandwidth transmission media makes multiplexing essential for efficient utilization of all the bandwidth available. *Time Division Multiplexing* (TDM), now the most widely employed multiplexing mechanism in digital transmissions, is easier to implement and cheaper to run than its predecessor.

In brief, here is how TDM systems work. Connecting exchanges allocate time slots between them to certain channels (see Figure 1-12). Only during that time slot can a switch transmit data from the allocated channel. Hence, information belonging to the first channel is transmitted in the first time slot. The second time slot is assigned to the next channel, and so on. When all channels have had a time slot—an opportunity to transmit—the cycle begins over with the first channel. The receiving side is synchronized with the sender side, so it can mark the end of one time slot and the beginning of another.

TDM too requires the use of buffers in the exchanges. An exchange has to store the information ahead of transmission in anticipation of the proper time slot's arrival. This, as explained previously, adds some small, but not onerous, delay to the switching process.

Note that the predominant transmission technique for telephony provides 64-kbps channels, exactly the same data rate as the output of PCM codecs. This makes these channels perfectly suitable for the digital transmission of PCM-encoded voice.

Digital Signalling Systems

The appearance of digital transmission made upheaval in signalling systems almost inevitable. Consider the fact that analog systems use tones and

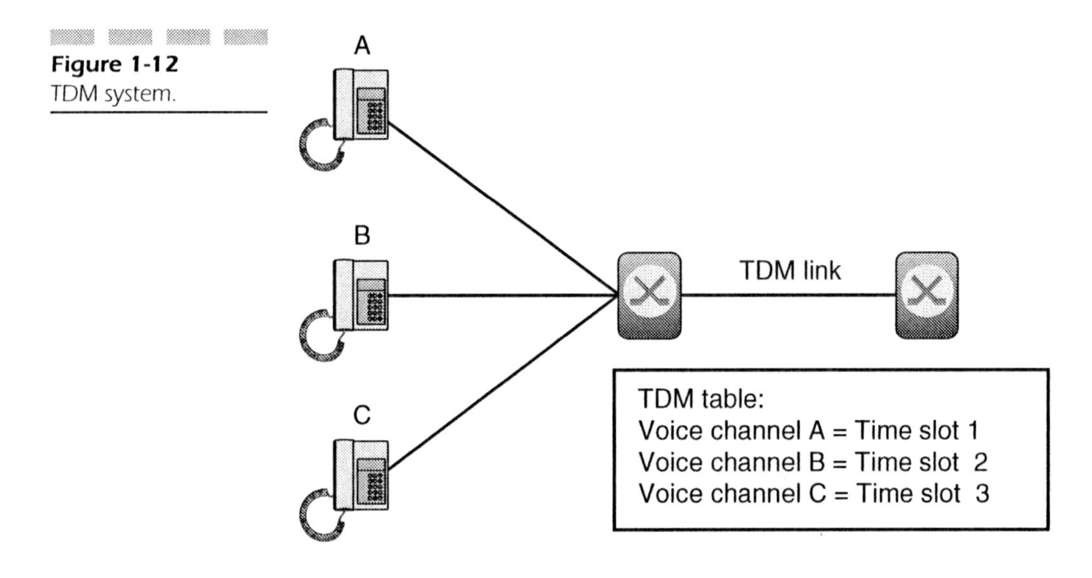

Figure 1-12
TDM system.

pulses for signalling. A pulse occupying a certain frequency means seizure of circuit and tones are usually employed for transmitting addressing information. The amount of signalling information that can be transmitted this way is quite low. Therefore, analog signalling systems were not only expensive but also inefficient exchange-to-exchange. Digital transmission, of course, *increased* the amount of information to be carried by the control plane. Digital signalling enabled switches to exchange bit streams in place of discrete pulses. Just as combining letters according to certain syntactic rules can create first words and then sentences, defining rules for bit streaming can create signalling messages. A signalling message contains much richer information than a pulse or a tone can. Signalling protocols codify the rules to form messages from bit streams and their semantics. It is protocols that spell out the actions to be taken under any circumstances known to prevail within exchanges. These actions are usually described by means of protocol state machines, so state machines can be thought of as the set of actions triggered by a certain event in a certain state.

Access and Trunk Signalling The first digital signalling protocols were implemented between exchanges. But the subscriber line remained analog, and thus so did the signalling procedures, leading to a disjunction between access signalling—signalling that passes between terminals and the network—and network signalling, also known as trunk signalling (see Figure 1-13). Later on, when the subscriber line became digital, this distinction between access and trunk persisted, and different protocols were used for access and for trunk signalling. The motivation was to permit complex exchanges but keep terminals relatively simple, enabling the network operator to control almost everything happening in the network.

The interface between terminals and the network is called *user-to-network interface* (UNI), and the interface between exchanges is knows as *network-to-network interface* (NNI). Protocols used in the latter are more complex because more information flows between switches than flows to the terminals. Terminals are only informed about what is essential in order to keep the user updated about the status of a call. This paradigm, consisting of an intelligent network and dumb terminals, opposes the IP paradigm in general, and the SIP paradigm in particular.

Access Signalling

What kind of control information is signalled in analoge subscriber lines? It consists of off/on hook signals, dialed numbers, and various tones that inform the user about call status. Off/on hook signals are used to request dial tone for a call attempt and to accept and terminate calls by picking up the phone and by hanging up respectively. The dialed numbers are transmitted from the terminal to the local exchange through pulses or tones. (Old systems employ decadic pulses, incidentally, while newer terminals implement *Dual Tone Multi-Frequency* [DTMF] tones.) Tones also convey status information to the user plane. Standard tones have been defined (such as busy tone or alerting), but they sometimes vary from country to country.

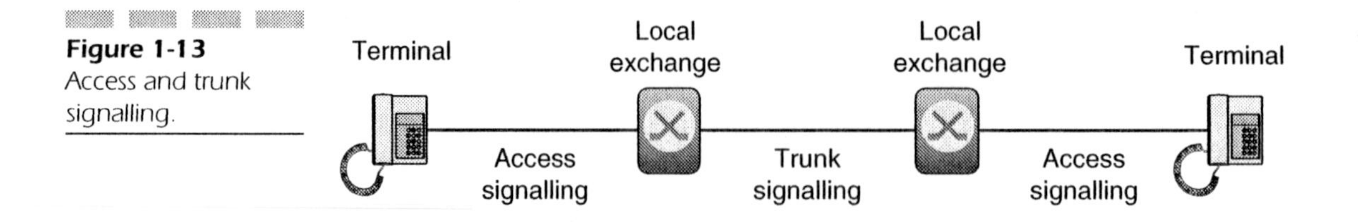

Figure 1-13 Access and trunk signalling.

The implementation of digital terminals made it necessary to develop digital access signalling systems. *Digital Subscriber Line No. 1* (DSS-1) is the most widespread. It is used in the ISDN and also in mobile networks; signalling between *Global System for Mobile* (GSM) communications mobile terminals and *Base Transceiver Stations* (BTS) is based on DSS-1. Figure 1-14 is an example of a DSS-1 call flow. It shows the message exchange between a terminal receiving a call and its local exchange.

DSS-1 is definitely richer than any analog signalling system but not as rich as commonly used trunk signalling systems, which we will look at next.

Trunk Signalling

Interexchange signalling in circuit-switched networks is referred to as trunk signalling. Two types exist: *Channel Associated Signalling* (CAS) and *Common Channel Signalling* (CCS). CAS systems preceded CCS systems, which offer better capacity, reliability, and flexibility, and at present are progressively being replaced by them.

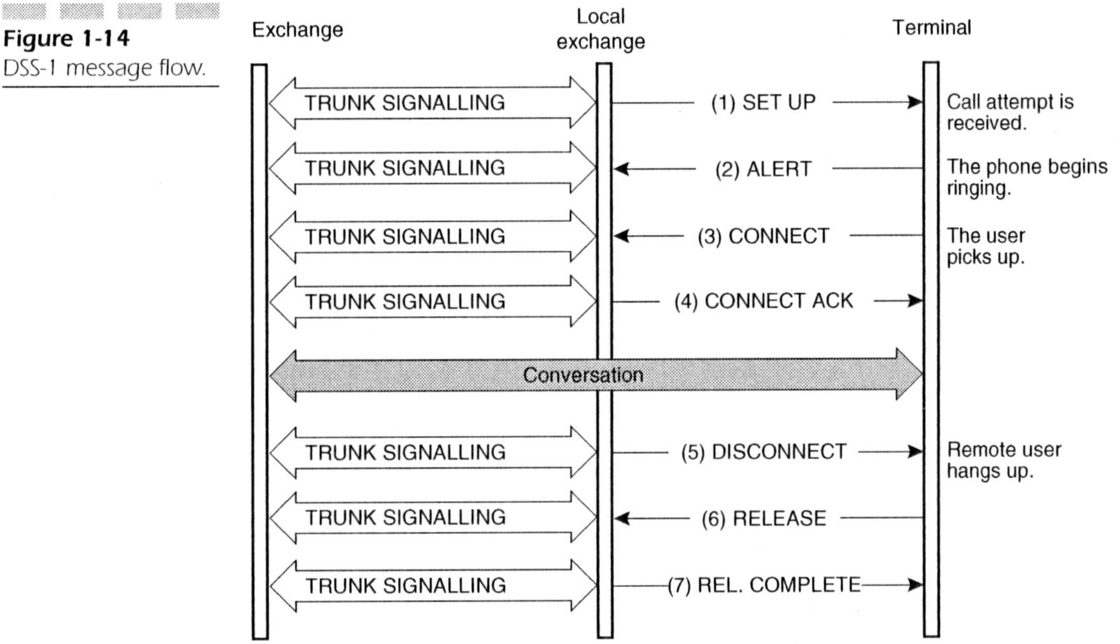

Figure 1-14
DSS-1 message flow.

Channel Associated Signalling In CAS systems, the signalling and the voice associated with a certain call are transferred along the same path. In-band signalling, which was described previously, is an example of CAS.

CAS can be used in both analog and digital systems. Digital systems that use CAS implement a specific channel for line signalling. Line signalling controls the establishment and release of voice channels (see Figure 1-15). The line signalling associated with all voice channels in a trunk is transported in a single channel. European systems, for instance, reserve time slot number 16 in each PCM link. Thus, the sixteenth time slot carries line signalling associated with channels 1 through 15 and 17 through 31 (time slot 0 is used for frame synchronization). Some examples of line signalling are: idle line, seizure of line, answer, and charging pulses.

On the other hand, register signalling, which handles addressing information, is transported through the voice path. Some examples of register signals are: callee's number, callee's status, and caller's number. Since register signals are sent only during call set-up, when there is still no voice transfer, they do not interfere with the user plane.

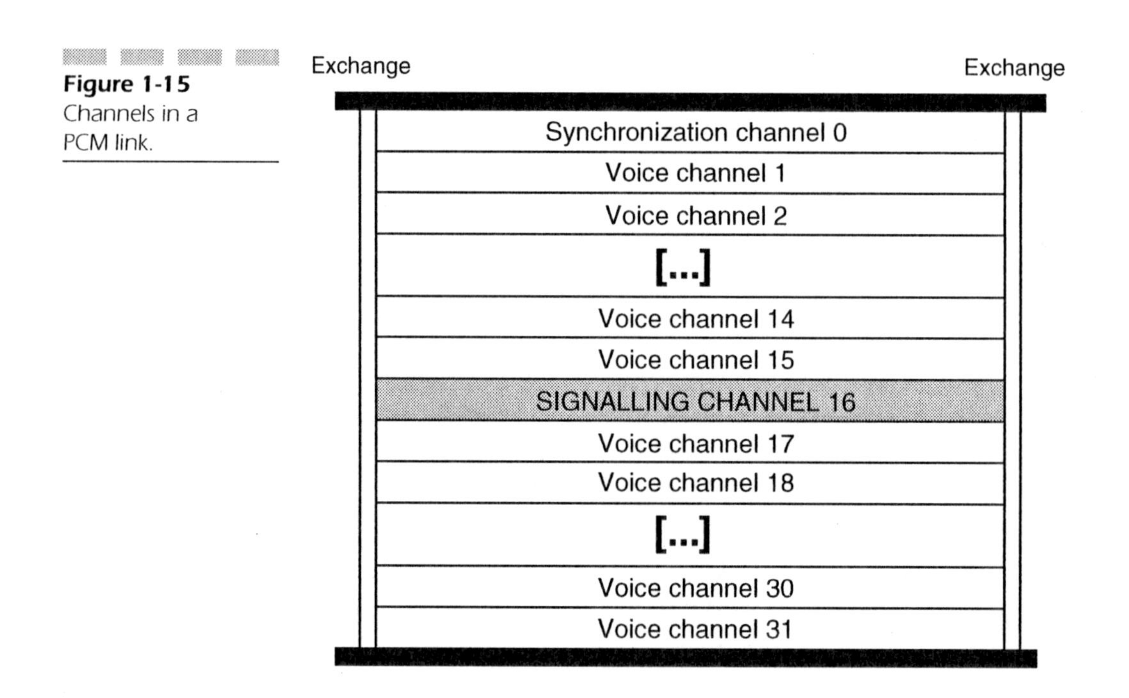

Figure 1-15 Channels in a PCM link.

Common Channel Signalling (CCS) In CCS, voice and signalling, as a rule, traverse different paths through the network (although their paths are still related to a certain extent). All the nodes handling media also handle signalling, so that, if a voice path traverses two exchanges, these two exchanges will also receive the associated signalling. How signalling is routed between them, however, is different in CCS than other systems. After leaving one exchange in the voice path, signalling can traverse a set of intermediate nodes, none of which handle voice before reaching the next one. We call these intermediate nodes *Signalling Transfer Points* (STPs).

STPs are a special feature of CSS, necessitating the implementation of a dedicated network optimized for signalling transport, and thus capable of better performance and reliability than shared voice/signal networks.

Dedicated Signalling Network Signalling traffic associated with a call is bursty in nature. It is very intense in certain phases of the call, like call establishment or call release, but otherwise quite low once the call is established. Bursty is good in that it enables a single signalling channel to control several voice channels. In practice, one signalling channel can handle several thousands of voice time slots.

Implementing a signalling network independent from its voice counterpart enables service creation and makes interworking different networks easier. SS7 is the most widespread CCS system at present, and it is SS7 in particular that allows the creation of services that go beyond basic CAS

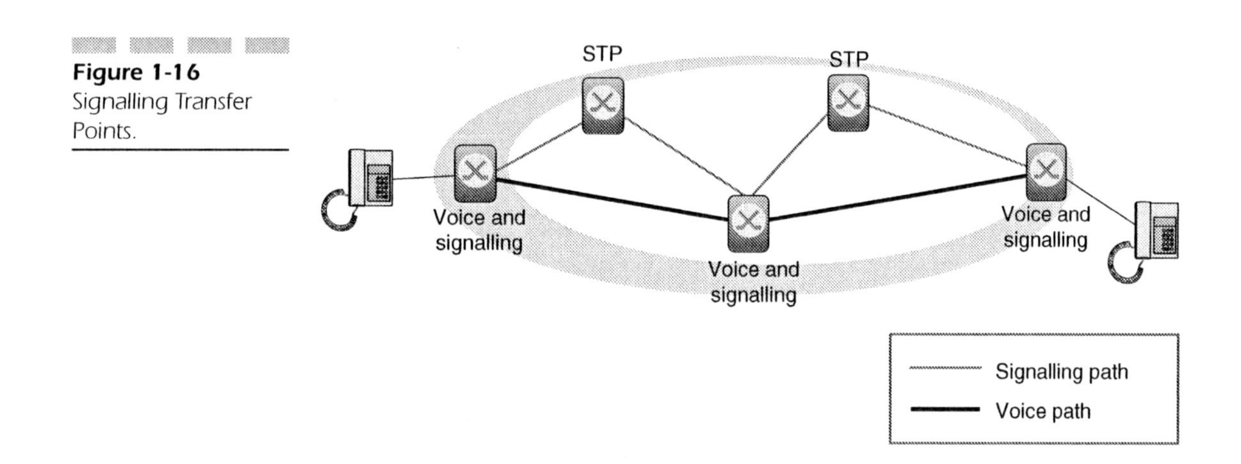

Figure 1-16 Signalling Transfer Points.

connections. The capability to signal between dedicated network nodes that do not handle voice is in fact the basis of all the advanced services provided by the current PSTN. By implementing service logic and switching logic separately, we encourage service creation and user plane manipulation to evolve independently.

For instance, when someone dials a toll-free number, the system must be able to look up the phone number dialed by the user (it'll be a 1-800 number in the United States) and translate it into the (geographically determined) number of the company the user wants to reach. This database look up is performed entirely through signalling messages (see Figure 1-17).

Toll-free and other services can be implemented thanks to the separation between the signalling and the voice path. Otherwise, it would be necessary to route the voice path towards the database and then back from it in order to perform the number look-up, leading to a severe decrease in voice quality experienced by the caller.

Today, two main CCS systems are in use: SS6 and SS7. SS6 was standardized in 1968 and was meant to be used in international connections. SS7 was standardized in 1980 and remains the dominant signalling system at present.

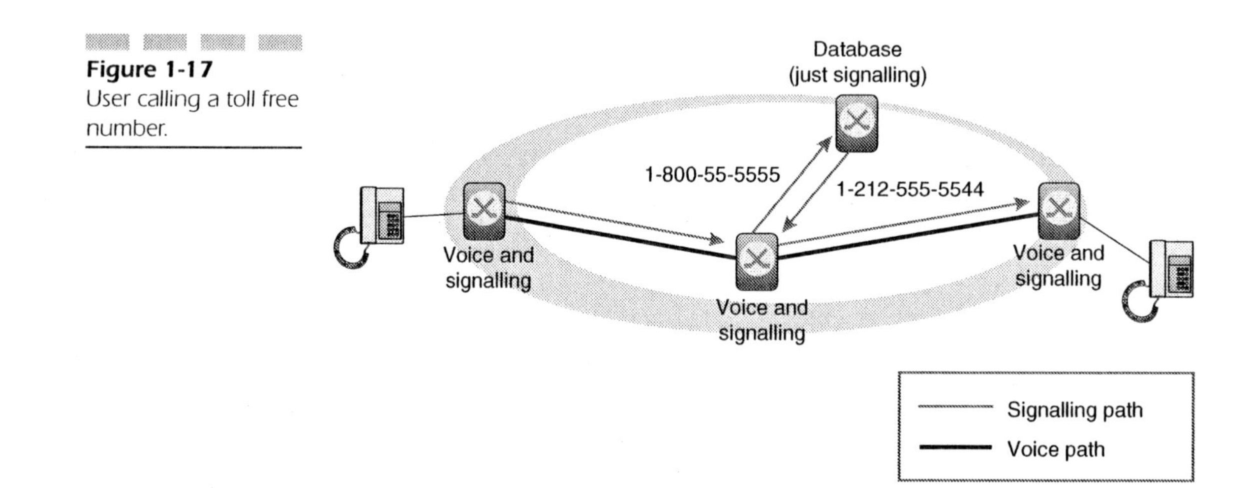

Figure 1-17
User calling a toll free number.

SS7

SS7 provides both circuit-related and non-circuit-related signalling. Circuit-related signalling is, as its name suggests, associated with the establishment of a voice channel. But frequently no voice channel has to be established—as when the system consults a routing table—and this kind of query can be handled by non-circuit-related signalling. As an enabler, non-circuit signalling was a vital step toward mobility.

Let's first look at circuit related signalling. *ISDN User Part* (ISUP) and *Telephone User Part* (TUP) are the relevant SS7 protocols in this regard. In the last specifications, ISUP included all the functionality supported by TUP, plus more features such as the capability to transfer user-to-user information, so ISUP is sufficient for our discussion here. ISUP is actually a generic name for a protocol which has many flavors in many different countries. When carriers want to implement a new feature, they often use non-compatible extensions that cause interoperability problems, and in so doing, create a variety of local ISUP flavors. Naturally, national variants of ISUP are incompatible, so an international version of ISUP is used to interconnect networks with ISUP variants. Gateway switches between them understand/support both international ISUP and the specific ISUP implemented in the network and perform translations between them.

The drawback of this arrangement is that international ISUP is comparatively feature-poor, limiting network services to users connected to the same network. For example, users in Spain must choose among an available set of services for calls within Spain, because the signalling protocol used for those calls is the Spanish ISUP. Users in Brazil have another set of services available for their internal calls. But services can't extend to calls between both countries, because there's no feature transparency between both flavors.

Supplementary Services Functionality provided by ISUP can be divided into three groups: basic services, circuit management, and supplementary services. Basic services include the basic call establishment to provide voice service.

ISUP circuit management functions include blocking and unblocking circuits and establishing, releasing, and testing circuit-switched paths. The same protocol used for establishing calls is also used for managing network resources.

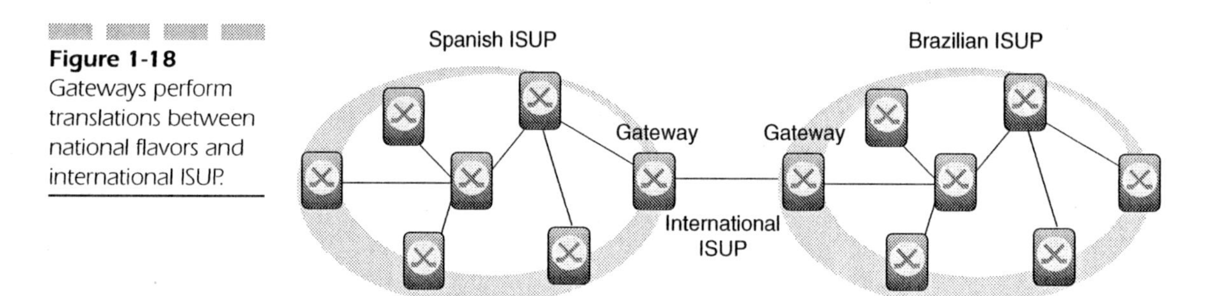

Figure 1-18 Gateways perform translations between national flavors and international ISUP.

Supplementary services are more varied; examples are call waiting, call transfer, and conferencing. These supplementary services are undeniably useful and many users take advantage of them. However, gaining access to them is pretty complicated. Because the interface used for this purpose is usually the keypad of a telephone set, the user has to learn different codes for different services. If the service is not accessed very often, remembering codes becomes difficult. A non-atypical example of service invocation is typing "*22# telephone number #" in order to turn on call forwarding, complicated enough that it will probably require the user to look up procedures before nearly every forward. We will see that the introduction of better user interfaces not only makes for more frequent usage, but also makes protocols evolve more rapidly, since the users can use newly introduced features in a more friendly way.

ISUP supplementary services are implemented in a distributed manner. Users gain access to new services when the local exchange to which they are connected is upgraded to support new features. Therefore, service availability depends on the local exchange and its business plan. Moreover, not all users of the network have access to the same services since all the local exchanges of the network cannot be upgraded at the same time and with the same version of software and the same hardware. Even users accessing the same service might expect different performance behavior, depending on which switch they are requesting the service from.

Intelligent Network (IN) Services These limitations of ISUP with respect to service provision prompted implementation of the *Intelligent Network* (IN) services, which utilizes *Intelligent Network Application Protocol* (INAP). Although ISUP services were distributed, IN services are clustered in a centrally located node. This node, called *Service Control Point* (SCP), can be accessed by any user in the network, so providers can deliver new services simply by upgrading the node—a big advance in service management. Services are implemented in the SCP with scripts—a set of rules and instructions that contain the particular service logic. Scripts come in two types: system and group. The system script performs number analysis in response to the number entered by a user, and then calls the proper group script to execute the actual service logic.

Distributed throughout the network are a number of IN service triggers called *Service Switching Points* (SSPs). An SSP analyzes the number entered by the user and decides whether intelligent network services are needed. It then contacts the SCP, which in turn determines the actions required to provision the service invoked.

The Paradigm Behind SS7

We've already touched on how signalling in the PSTN has evolved from its analog origins advanced. Now let's analyze the architectural principles behind the telephone network and how they differ from those used to design the Internet. This comparison will go a long way toward explaining why SIP is a true revolution, and not an evolution of existing signalling systems.

Two main architectural differences exist between the PSTN and the Internet:

- Intelligence in the PSTN is concentrated in the network rather than the terminals.
- Access to the network and the services provided are tightly related.

Intelligence Inside the Network SS7 makes a clear distinction between terminals and the network. It operates in the network—trunk signalling—while other protocols like DSS-1 bridge the network and its terminals. SS7 proposes an architecture that enables the use of very simple terminals with limited responsibilities. On the Internet, the opposite holds true. Intelligence is pushed to end-user equipment, and the network is kept as simple as possible. We will see in a later chapter that this enables still faster and more flexible implementations of new services.

Following out the logic of the Internet paradigm, a signalling protocol should support intelligent end systems instead of building a network that makes decisions on the user's behalf. As we have seen, protocols in the PSTN do not meet this criterion. Some systems using SS7 such as GSM resemble master/slave architectures, where a node in the network—the *Mobile Switching Center* (MSC)—sends commands to the GSM terminal, which obeys unconditionally. This architecture leads to a network that makes assumptions about what the user wants, instead of letting the user dictate what he wants. When end-to-end services are implemented in such a network, unintended effects sometimes occur.

Consider, for instance, a call between two GSM terminals (see Figure 1-19). GSM terminals use GSM codecs for encoding the voice. The MSC that handles the caller's terminal decides, on behalf of the terminal, that transcoding from the GSM codec to PCM—the codec used in the fixed PSTN—is needed. When the call reaches the MSC handling the callee's terminal, it performs the opposite transcoding, from PCM to GSM, before at last sending the voice to the terminal. Unhappily transcoding notably reduces the voice quality, and when it is performed twice, as in this scenario, its effect is amplified.

This example shows clearly that end-to-end services are better implemented using end-to-end protocols, which would enable both terminals to negotiate the codec to be used, avoiding the situation we've just examined.

Reliability Because the PSTN concentrates system intelligence in some network nodes, a failure in one of the nodes means disruption of service for the user. Consequently, network nodes must be highly reliable. Techniques such as processor redundancy ensure that a certain network node does not just

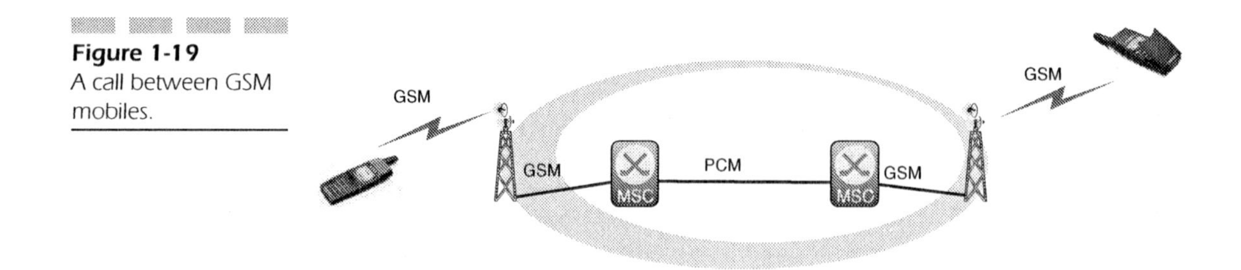

Figure 1-19 A call between GSM mobiles.

fail. However, implementing nodes that (almost) never fail results in a significant increase in equipment costs.

We will see that on the Internet, since the intelligence is pushed to the terminals, failure in a network node such as a router is usually not critical. Other nodes can take over its task. Here reliability is obtained by implementing end-to-end services in which a network failure won't lead to a loss of state information essential for the service. As long as the end systems do not crash, the service can still be provided.

Access Tightly Tied to Service Provision We have seen how PSTN signalling has evolved throughout the years, and why its evolution has been closely related to that of the user plane. When new multiplexing techniques or transmission mechanisms are created for the user plane, signalling protocols have to evolve to exploit the new mechanisms. Traditionally, new signalling protocols were developed in order to carry new information about the user plane.

Signalling protocols specify the types of user plane that can be used in conjunction with them in detail. For instance, international ISUP defines three types of user plane: 64-kbps unrestricted, speech, and 3.1-kHz audio.

This is because, in the PSTN, access to the network and service provision is coordinated. Therefore, if a user accesses the PSTN, the services he or she is able to access will be those provided by the protocol used for access. If DSS-1 is used, for instance, the caller will be provided with a voice channel. He or she cannot then access the PSTN and request another kind of service.

In the next chapter, we'll see that on the Internet environment, access and service provision are completely separated. A user can (and often does) connect to the Internet through a particular *Internet Service Provider* (ISP) and get e-mail services through an Internet portal. Access provider and service provider in this situation are different.

As one might expect, the Internet promotes a different approach to design signalling protocols from the PSTN. We will see that SIP does *not* restrict the number of types of sessions that can be established. A caller can use SIP to establish a VoIP session, but he can also use SIP in order to establish, say, a gaming session. SIP could even be leveraged to establish a traditional telephone call. Thus, Internet protocols are implemented in a much more modular way than PSTN protocols.

Conclusions

We have seen that the PSTN implements the intelligence of the system in the network rather than in the terminals. In addition, signalling related to call establishment is tightly tied to management of the user plane. This overloads protocols and makes them less general for use in another type of network.

Taking the previous statements into consideration, we infer that the paradigm behind SS7 does not suit IP networks, which are designed for maximum flexibility. SS7 is an excellent performer in circuit-switched networks but cannot take advantage of the very flexibility that IP networks provide. Therefore, signalling protocols to be used in IP environments need to be designed keeping the IP paradigm in mind. It will be seen that SIP suits it very well.

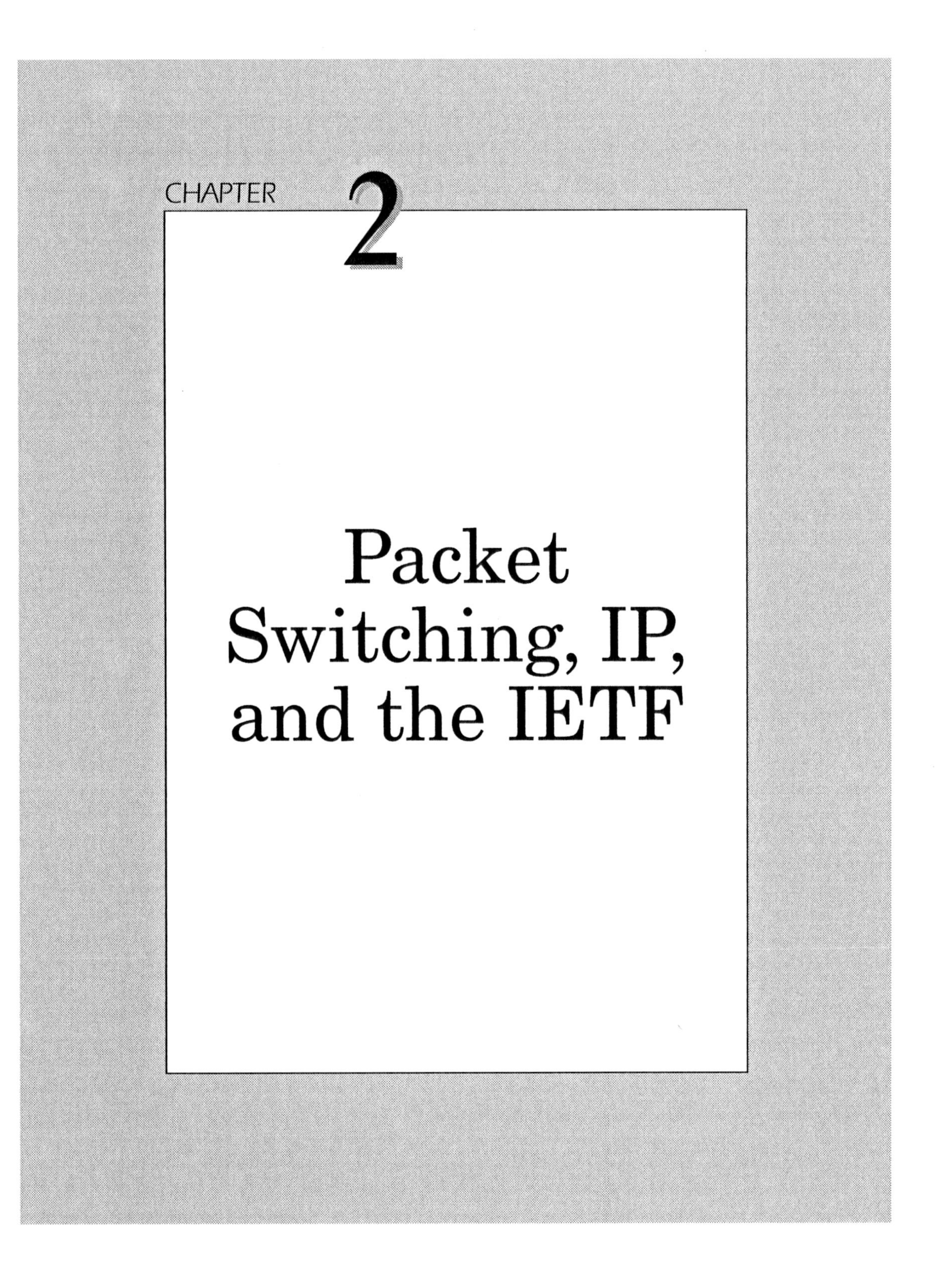

CHAPTER 2

Packet Switching, IP, and the IETF

In this chapter, we will learn how the Internet is built, how packet switching works, and how Internet protocols are designed. This information is vital context for understanding the motivation behind the way the *Session Initiation Protocol* (SIP) works.

In addition, a familiarity with how the *Internet Engineering Task Force* (IETF) works emphasizes how mature SIP and its different extensions really are. It is important for basic IP literacy to know the process that every Internet protocol follows before becoming an Internet standard.

Packet Switching

We saw in Chapter 1, "Signalling Protocols in the Circuit-Switched Network," that circuit switching was originally developed for analog transmissions. Circuit switching provides a dedicated path from source to destination that is suitable for transmitting continuous analog signals such as voice. *Frequency Division Multiplexing* (FDM) was the most widespread multiplexing technique in circuit-switched analog networks.

The appearance of digital technology enabled circuit switching to evolve. Information is transmitted between systems in binary format using 0's and 1's. Network nodes can easily store digital data and resend it without degrading the quality of the data. This capability enables the use of multiplexing techniques that are more efficient than FDM, such as *Time Division Multiplexing* (TDM).

However, even TDM systems make rather inefficient use of the network resources. Voice transmissions fail to use all the available bandwidth in a circuit, none of which can be reassigned for any other purpose.

As if that weren't enough, data communications brought new communications patterns into the picture. Data traffic is bursty and non-uniform. Terminals do not transmit continuously, but are idle most of the time and very busy at certain points. Data rates are not kept constant through the duration of the connection either; they vary dynamically. A particular data transmission has a peak data rate and an average data rate associated to it, and these are not usually the same.

Employing dedicated circuits to transmit traffic with these characteristics is a waste of resources, as we can see in Figure 2-1. The circuit in the

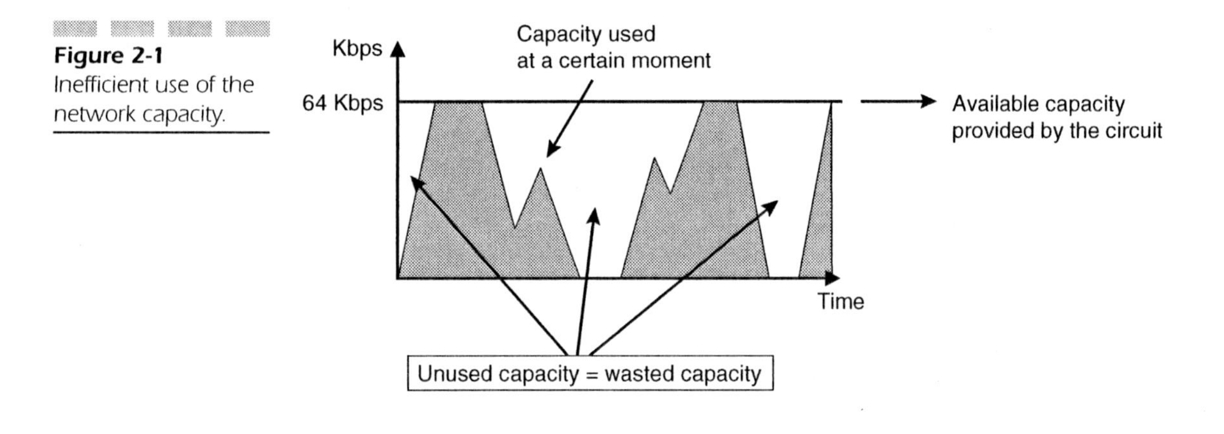

Figure 2-1
Inefficient use of the network capacity.

figure provides 64 Kbps continuously, but the data rate needed by the systems communicating is rarely 64 Kbps; it is usually less and sometimes more. The figure also shows that the available capacity that is not in use by the system at a certain moment is simply lost.

Packet switching was first designed to fulfill the requirements of bursty traffic presented by data. Although the first papers about packet-switching technology appeared in 1961, the first packet-switching node was not implemented until the end of the 1960s. In packet switching, the content of the transmitted data determines how network switching is performed. The source nodes chunk the information that will be transmitted into pieces called *packets*. Then the information needed to route each packet to its destination (such as the destination address) is appended to the packet, forming the *packet header*. The packet is sent to the first network node; in packet switching, network nodes are referred to as *routers*. Figure 2-2 illustrates a network of routers. When the router receives the packet, it examines the header and forwards the packet to the next appropriate router. This process of looking up the header of the packet is performed in every router in the path until the packet reaches its destination. After reaching the destination, the destination terminal strips off the header of the packet and obtains the actual data that was originated at the source.

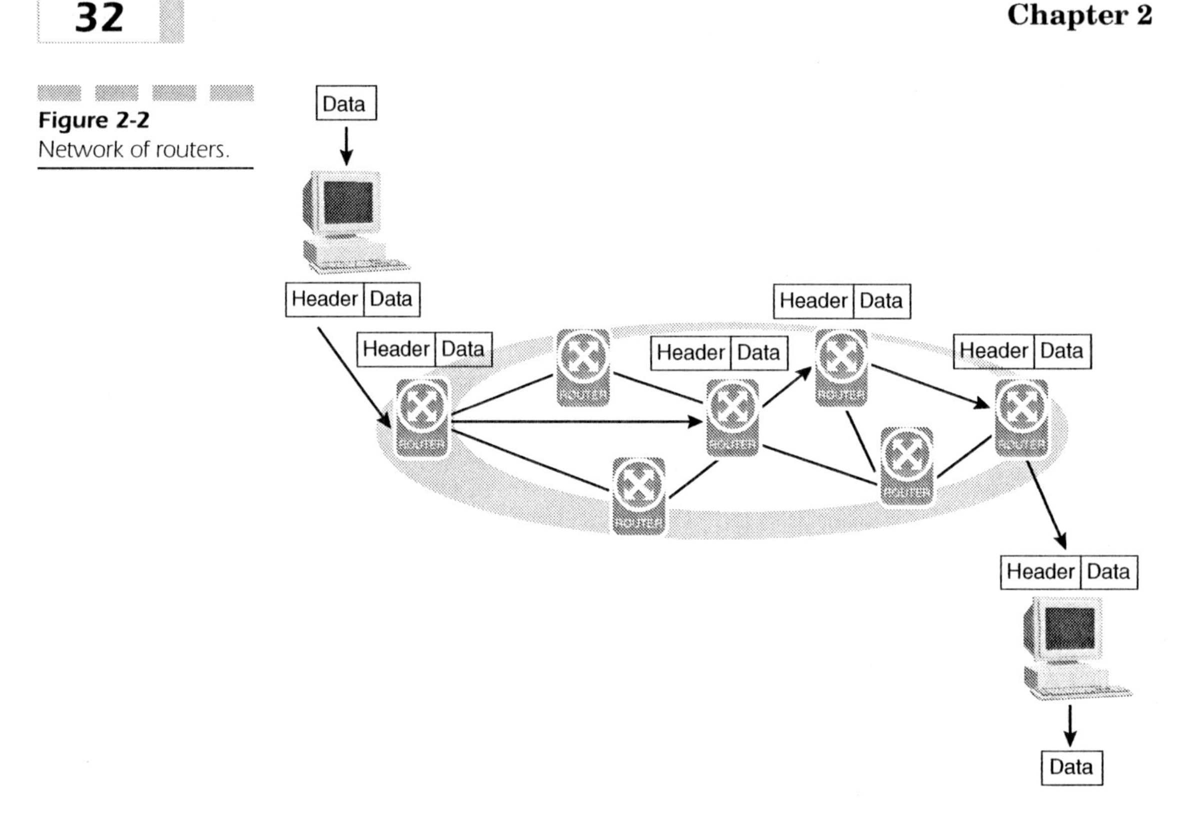

Figure 2-2 Network of routers.

Datagrams Routing in packet-switched networks can be performed with datagrams or virtual circuits. Networks using datagrams make routing decisions based solely on the header of the packet that will be routed. The header is examined and the packet is forwarded to the next proper hop. Different packets addressed to the same destination might take different routes through the network, enabling *load balancing.* Packets can avoid congested points in the network by taking alternative paths. As a consequence, packets taking different routes will experience different delays, and it's entirely possible for packet #5, which was sent later than packet #1, to take a faster route and arrive earlier than packet #1. Out-of-sequence arrival of datagrams is remedied by the destination, which reorders them before delivering them to the user.

Figure 2-3 shows how the datagram approach works in the network. Datagrams traveling from A to B are routed towards B based on the header of each particular datagram. We've already recognized that a particular router might decide to route datagrams towards B using alternative paths. The result of this approach is that datagrams reaching B have traversed different paths and therefore might have experienced different delays.

Virtual Circuits Virtual circuits behave more like circuit switching. Routing decisions are made based on the header of the packet that will be routed and on the previous packets that traversed the router.

The source, prior to any data transmission, sends a virtual-circuit establishment request to the network. The network then routes the packet containing the request towards the destination and assigns a virtual-circuit identifier to the path. Upon reception, the destination confirms the establishment of the virtual circuit by sending an acknowledgement packet to the source. Once the virtual circuit is established, all the packets carrying the data that will be transmitted will be routed through the same path.

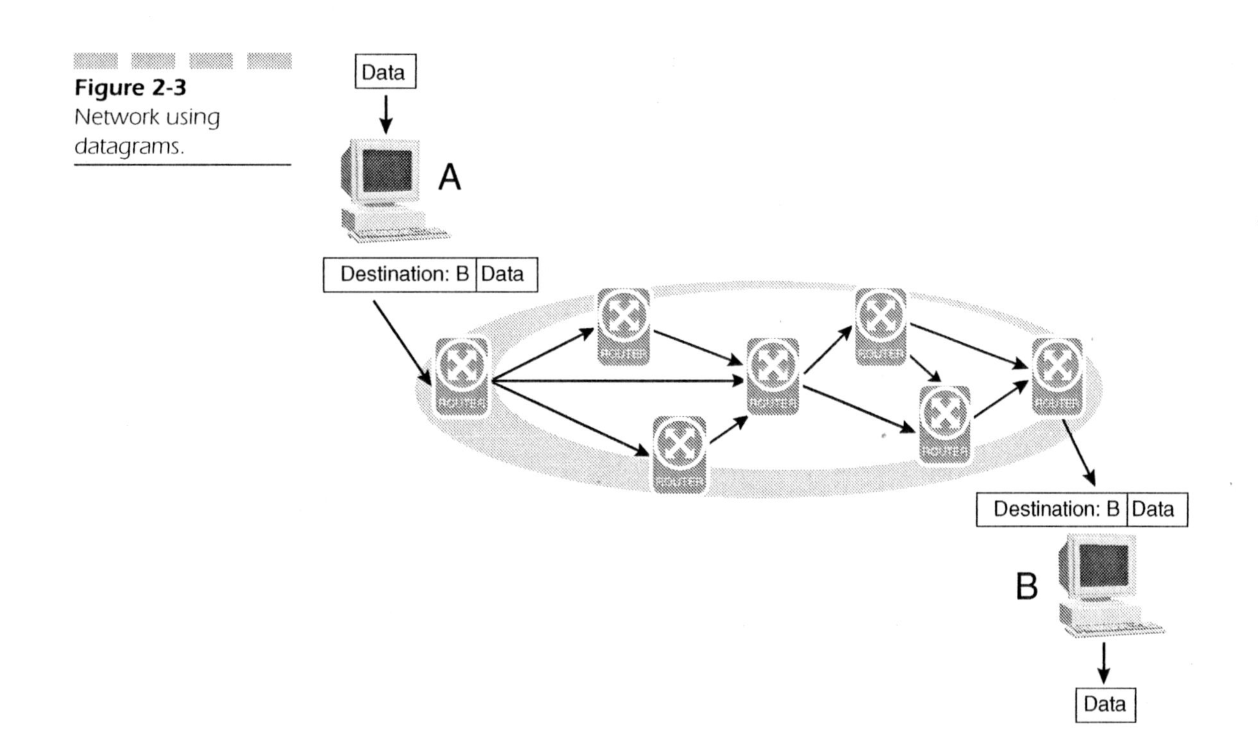

Figure 2-3 Network using datagrams.

Packet headers on the path incorporate the virtual-circuit identifier that was assigned previously, and routers forward packets accordingly.

Figure 2-4 shows a network based on virtual circuits forwarding packets from A to B. All packets follow the path of the first packet.

Differences Between Datagrams and Virtual Circuits Perhaps the most important difference between datagrams and virtual circuits is the state that each requires the network to store. Datagram routing requires no state information (besides routing tables) inside the network. Networks handling datagrams are stateless in the sense that once a routing decision is made, the router does not store the result in order to make subsequent decisions. Conversely, networks dealing with virtual circuits must store the state of virtual circuits and identifiers to match incoming packets to the proper vitual circuit. Therefore, virtual-circuit networks implement more intelligence in the network than datagram networks.

Each of these approaches to packet routing presents advantages and disadvantages. Datagrams do not require any establishment time prior to the

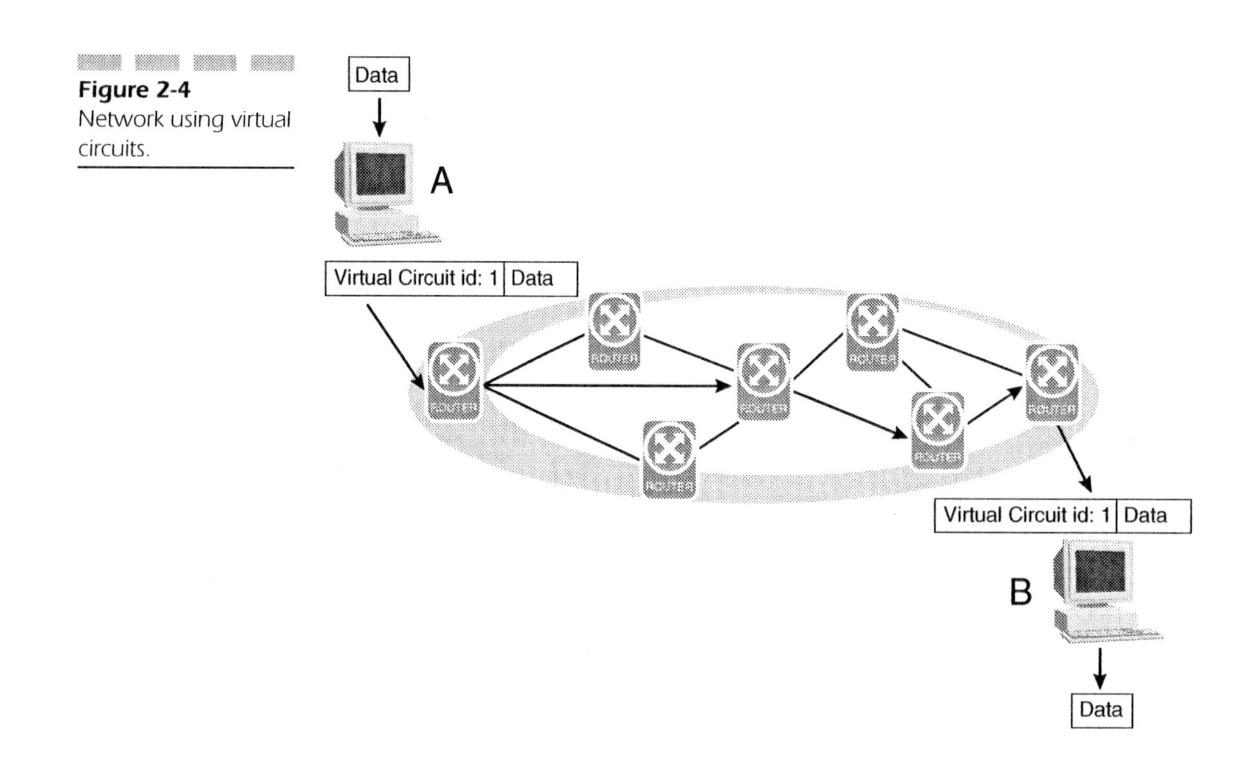

Figure 2-4 Network using virtual circuits.

transfer of user data, whereas virtual circuits have to be requested and established before any user data can transmit. When a terminal wishes to send packets to a number of different destinations, this establishment delay is present every time a new destination address is used. In such situations, datagrams provide more flexibility.

On the other hand, once a virtual circuit is established, it usually provides lower overhead than datagrams because headers have less to carry; they just carry a virtual-circuit identifier. Thus, for long transmissions, networks using virtual circuits are a better choice.

Datagrams take advantage of dynamic routing in the network. If a node in the network fails or the network becomes congested, packets can immediately take alternative routes to avoid stress points, without needing to establish a new circuit. Load balancing is also enabled; datagrams headed for the same destination can take different paths to distribute the network load more evenly. Virtual circuits can handle some load balancing, but when packets always take the same path, redistribution becomes more complicated.

Strengths of Packet Switching

Efficiency and suitability are the reasons why packet switching compares favorably to circuit switching. It returns unused resources promptly to the network. Packet switching also provides data rate adaptation between different terminals. Routers are designed to buffer incoming packets until they can be sent on the outgoing interface, so the incoming data rate is not necessarily the same as the outgoing data rate. This enables terminals using completely different data rates to communicate. In effect, the network performs data rate conversion.

Price is another strength. In general, packet-switching equipment costs less than circuit-switching gear because it's simpler. A router is basically a computer with several network interfaces attached to it. It receives packets from one interface, analyzes the headers, and forwards them through the proper outgoing interface. Falling computer prices in the last few years directly benefit packet-switching nodes.

Weaknesses of Packet Switching

We've cited more delay and higher overhead as the two primary reasons why packet switching may not perform as well as circuit switching under certain circumstances. Circuit-switching nodes do not examine contents of the information transferred at all, whereas packet-switching routers examine the headers of every packet before forwarding them. This process takes necessarily longer than, say, just switching information from one time slot to another in a TDM network. Besides, packets that will be forwarded through a certain interface in a router must line up in the outgoing queue of that interface. There, they wait until all of the previous packets are gone. When the network has a heavy load, router queues are usually full, which is compounded by the fact that packets have to queue in each and every node they traverse. It is possible to mitigate the delay introduced by implementing different priorities for different types of traffic in the queues, for instance, but these delays are inherent to packet switching and can't be eliminated altogether. How much of a disadvantage is the overhead introduced by full-packet headers? The headers, of course, consume network resources, and if the payload of a packet is small enough, the header can actually outweigh it. For instance, if a 20-octet header is used to transfer 5 octets of user data, the network will transmit a packet of 25 octets, in which just 5 octets are user data.

X.25

In the 1970s, the telephony carriers developed packet-switched networks based on virtual circuits. These networks offered many similarities to the *public-switched telephone network* (PSTN) and a smart network. The specification X.25 defined the interface between the terminals and the network (user-to-network interface), and X.75 defined interactions between nodes in the network (network-to-network interface). Networks implementing these specifications are colloquially known as X.25 networks.

X.25 networks provide high functionality to terminals. Network nodes perform error detection by exchanging acknowledgment messages to ensure that the information transmitted through the link has not been corrupted. Terminals only have to release the data, and the network performs flow and error control until the data is safely delivered. As we observed in the telephone network, a network providing high functionality enables the implementation of simple, reliable terminals to make use of it.

IP and the Internet Paradigm

X.25 networks present some weaknesses. Overloading the network with extraneous tasks such as flow control reduces its performance. Thus, packet-switched networks implemented using the PSTN paradigm did not exploit all the possibilities this technology offers. A paradigm shift was needed in order to make the best use of the newly developed technologies. This new paradigm, of course, is the Internet paradigm, and its main protocol is *Internet Protocol* (IP).

IP Connectivity

The Internet differs from other networks in that its sole purpose is providing connectivity. (The purpose of the PSTN, for instance, is providing telephone services and the purpose of the TV network is providing broadcasts.) A variety of services such as e-mail, the World Wide Web, videoconferencing, and file transfer are implemented based on end-to-end IP connectivity (Figure 2-5).

In order to achieve true end-to-end connectivity, a common end-to-end protocol is implemented at the network layer—the Internet Protocol (IP) [RFC 791]. The implementation of IP by all systems within the network has two advantages:

- The network is independent of the underlying technology.
- Applications can make use of a common IP infrastructure.

From a telecom perspective, IP infrastructure is quite a departure. Specifically, it enables us to connect networks that use different link-layer technologies to build a homogeneous network at the IP layer despite heterogeneous lower layer technologies (Figure 2-6). Another gain coming from

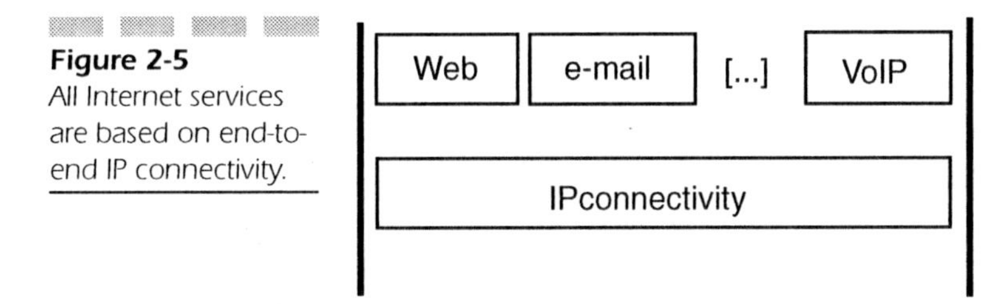

Figure 2-5 All Internet services are based on end-to-end IP connectivity.

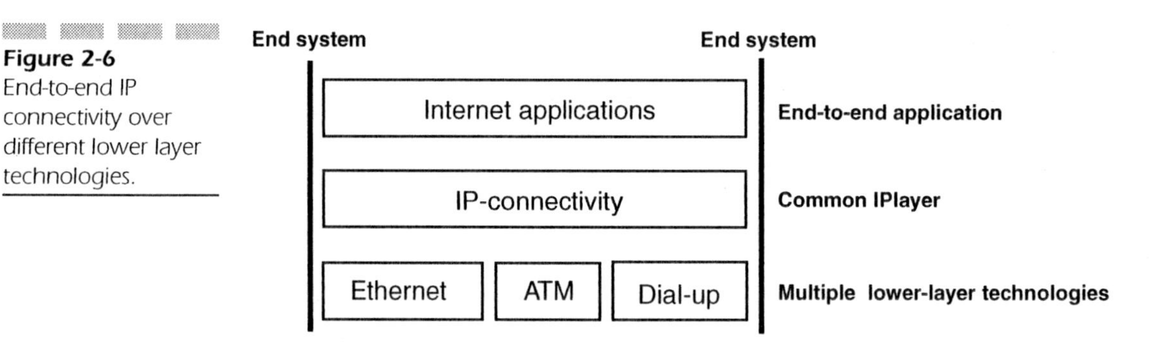

Figure 2-6 End-to-end IP connectivity over different lower layer technologies.

the use of a single protocol at the network layer is that any new lower layer technology can be used to transport IP traffic.

Currently, IP traffic travels on top of, among others, *Asynchronous Transfer Mode* (ATM), frame relay, leased circuits, optical fibers, and Ethernet *Local Area Networks* (LANs). In the future, IP will most likely run on top of newly developed high-speed digital technologies as a matter of course.

Once end-to-end connectivity is bestowed by the use of IP in every system of the network, IP becomes a common platform to develop applications and services. As with many protocols, developers must understand the ideas behind IP architectures before they can put them to good use, and those ideas invariably encompass a way of thinking that is worth examining.

Intelligence Pushed to the End Systems

An IP network consists of a set of intelligent hosts connected to a *dumb* network of routers that just provides datagram transmission—unreliably at that. IP is used at the network layer by both routers and hosts, relegating intelligence to the end systems. End systems are responsible for controlling IP network traffic, including end-to-end flow control and error detection. The router network is only required to undertake one simple task as efficiently as possible: the unreliable delivery of datagrams to the destination end system. End systems only know the behavior of the end-to-end traffic. The network does not send any notifications to the end systems indicating whether or not a packet has been delivered.

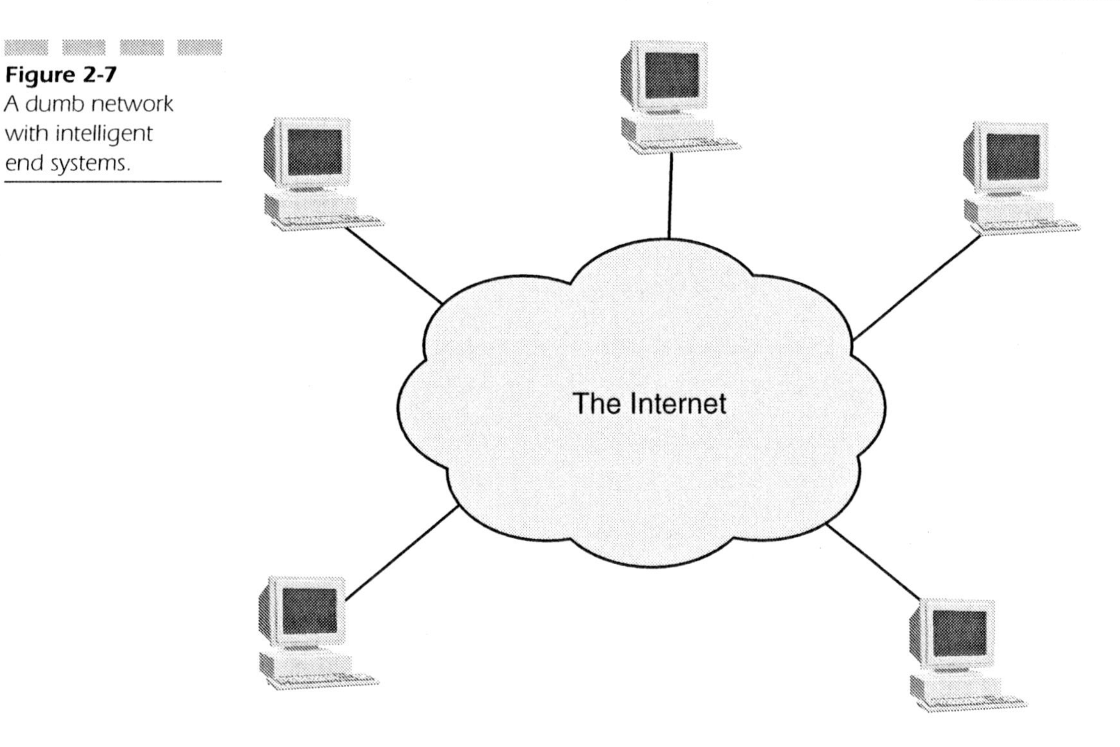

Figure 2-7
A dumb network with intelligent end systems.

Consequently, IP networks are almost stateless; they do not need state information in order to route packets towards their destination. This lack of state in the network makes node failures less dramatic because they do not store any state information necessary for end-to-end communication. This represents an interesting approach to engineering high availability. When some area of the network fails for any reason, the traffic between end systems is simply rerouted to avoid that area. Other routers can pick up the task of routing datagrams that were previously handled by the failed routers because no state information needs to be transferred between them for routing to resume. Thus, high availability is achieved by provisioning multiple paths rather than trying to implement fail-safe routers, which is a much harder job.

Figure 2-8 shows the chain of events when a router fails between A and B, and traffic between both nodes must be redirected. Failure in the router does not break the traffic flow between both end systems.

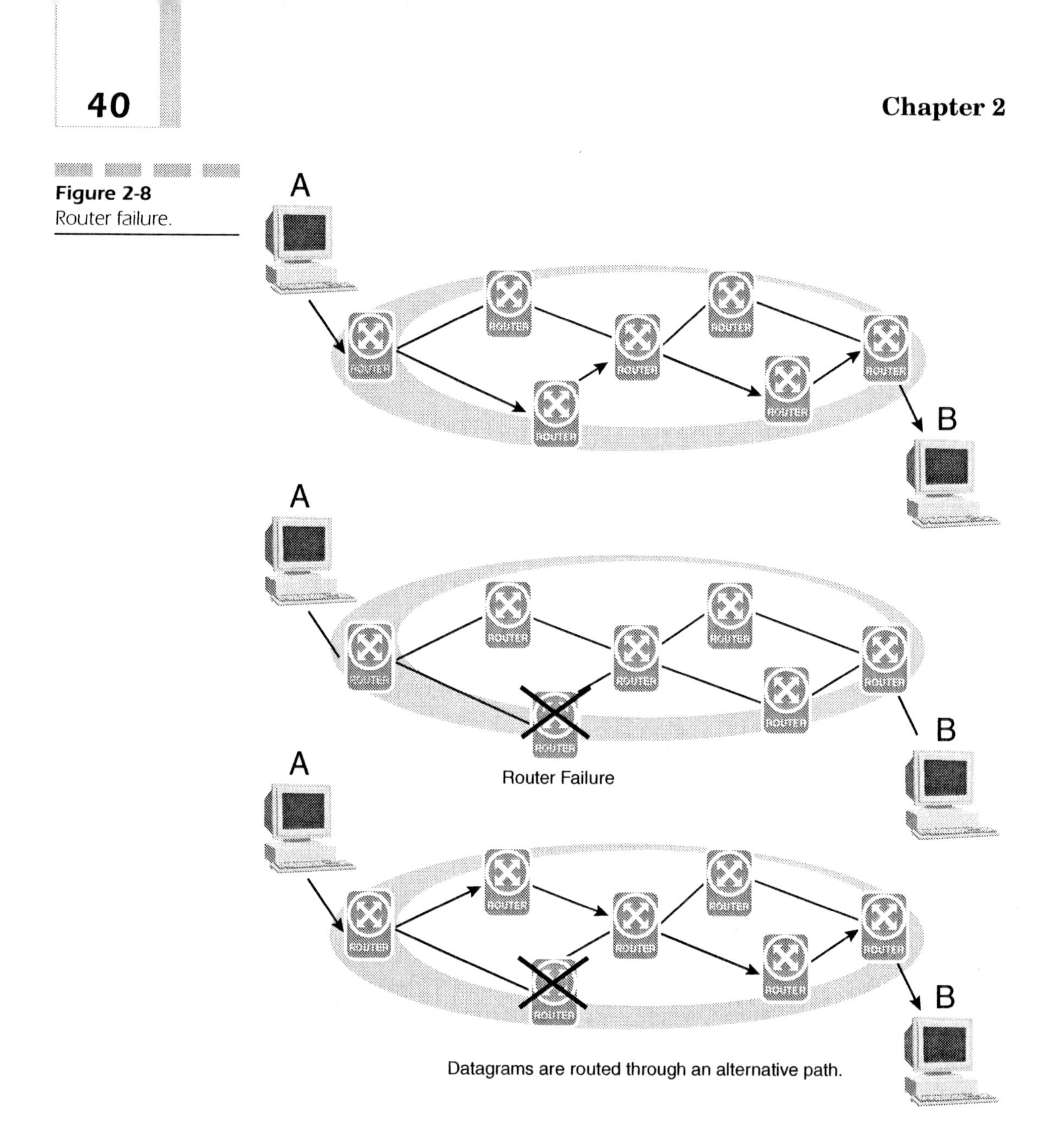

Figure 2-8 Router failure.

Note that although it is correct to state that IP networks do not store state, this doesn't imply that they are stateless. Some state is always needed in order to be able to route datagrams. For instance, when routing protocols distribute new routes to different destinations, the routers must store this information. Another example where state information is stored in the network is header compression in low-bandwidth point-to-point links. Both ends of the link store information in order to perform the header transfer. This happens, for instance, when encoded voice is transferred through a radio interface. In general, however, IP networks tend to keep the state information stored inside the network at minimum in order to scale better and to be more robust.

End-to-End Protocols

The presence of an IP layer common to all nodes and the concentration of intelligence in the end systems make IP an excellent platform for service creation, and a wide range of applications is already implemented on top of IP. Application development has as many approaches as practitioners, but in order to take full advantage of all the good features IP offers as a platform, developers are often advised to follow certain general design rules, which constitute the so-called IP way.

At this point, I want to emphasize that the importance of IP and the Internet (the largest worldwide IP network) is in part a function of its design rules and the way Internet standards are developed. IP is not just a network-layer protocol. IP is a completely new way of developing technology.

Perhaps the best-known rule of thumb is to try to mirror the behavior of IP by implementing intelligent end systems and a relatively dumb network. This practice encourages the design of end-to-end protocols at layers above IP, where they can deliver better end-to-end functions.

The terminal usually has the best knowledge in the system to provide the service that the user wishes to have. The network does not make unnecessary assumptions about what the user wants, but it provides transmission at a fairly predictable level. Another reason for using upper-layer end-to-end protocols is end-to-end security—security that doesn't depend on any (vulnerable or fallible) mechanism provided by the network. For applications that are secured this way, the information flowing between end systems is encrypted and/or signed by the end systems themselves. In the case of encrypted messages, the network cannot access the contents transmitted. In the case of digital signatures, nothing outside the end terminals can modify the messages. In both situations, end-to-end protocols are the only means to provide end-to-end security without impinging on the functionality required for each application.

General Design Issues

The IETF proffers a set of guidelines for protocol design in RFC 1958 [RFC 1958]. The express purpose of these guidelines is to improve the technical quality of the protocols documented in *Request For Comments* (RFCs) and to save time in the standardization process. If a protocol is properly designed from the beginning, the community does not need to give as much feedback to the authors and it can advance in the standards more efficiently.

The IETF Toolkit The design process followed by the IETF is usually referred to as a bottom-up approach in order to distinguish it from approaches that start by defining an architecture and then design more granular protocols between the nodes. The bottom-up approach provides protocols that solve specific needs of the community. A concrete protocol resolves a concrete problem, so all of these protocols taken together can be viewed as a toolkit, as shown in Figure 2-9. They can be (and often are) combined in different ways to resolve larger problems downstream.

The advantage of this approach is that a protocol becomes a component that can be used in several scenarios (see Figure 2-10), promoting a certain coherence in the RFCs. Thus, it is not necessary to design a new protocol every time we study a new scenario. When functionality is needed and no protocol provides it, a new protocol is in the offing; however, it needn't be a

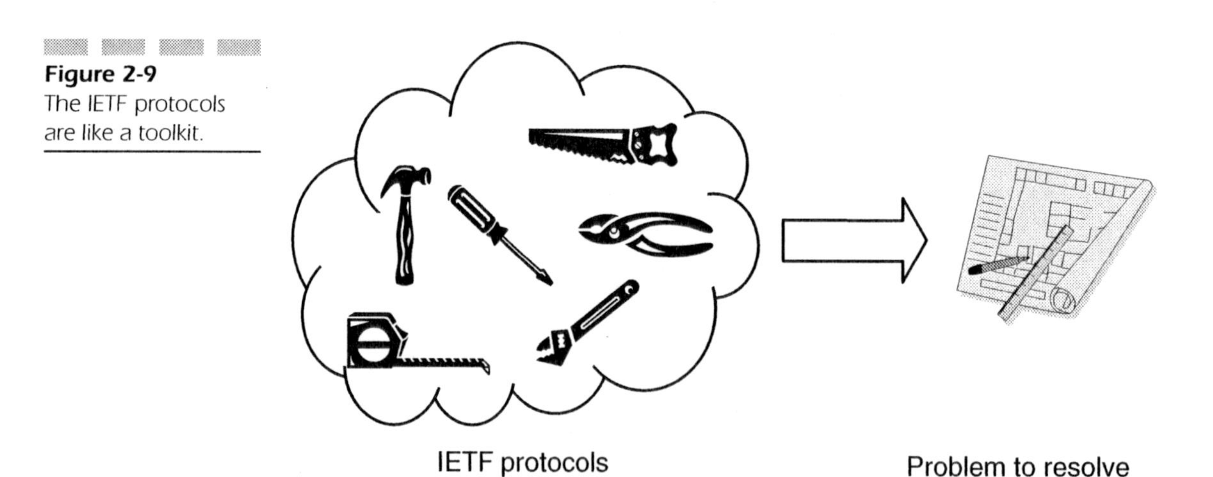

Figure 2-9 The IETF protocols are like a toolkit.

Figure 2-10 The same IETF protocol can be reused to resolve different problems.

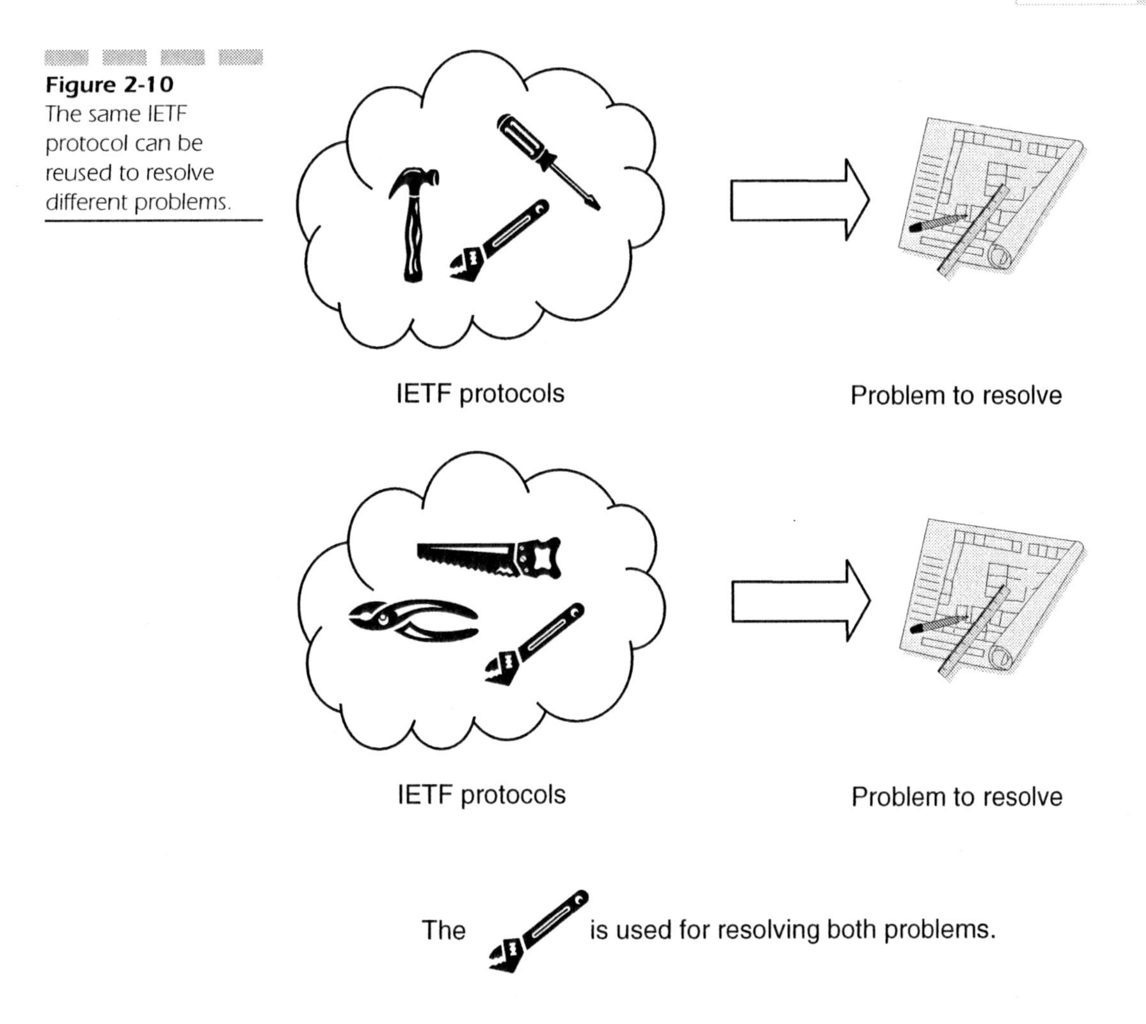

new solution. The ideal new protocol will address a general problem that will or can arise in other applications in the future.

For instance, Web servers have to authenticate the identity of users requesting access to a certain Web page. The IETF will automatically generate designs for a general authentication protocol usable by many kinds of servers, instead of coming up with the most direct solution for processing Web-server requests. In this way, they avoid duplication of protocol functionality as much as possible and add elegance to Internet engineering. The IETF designs protocols in a modular fashion.

IETF Protocol Features *Simplicity* is a key factor in protocol design. Simple solutions are always preferable to more complicated ones. Sometimes performance gains or increases in the functionality of a protocol come with a cost—more complexity. In certain situations, introducing a new level of complexity is unavoidable, but when a choice is available, the simpler solution is the rule.

Because all of the protocols coming from the IETF are used in the Internet, they must also meet stringent requirements for *scalability*. No matter what the scenario that originally provoked a new protocol might look like, the resulting protocol should be usable in large networks with several millions nodes.

Robustness is another feature every protocol should exhibit. The IETF approach to protocol robustness is well captured in one sentence: be strict when sending and tolerant when receiving. An implementation should always strive to follow correct protocol syntax and send well-formed messages. However, messages received by one implementation from another should not be scrutinized for grammatical errors or syntax problems. The guideline is that any message that can be understood should be processed, and adherence to that guideline enables correct implementations to interwork properly with faulty ones.

For instance, suppose a protocol specification states that a certain message containing a timestamp has to be sent every 500 ms. A merely *correct* implementation would send the message every 500 ms, but a *good* implementation would work just as well if it receives messages every 400 ms instead. Obviously, this guideline is not endlessly flexible; it's meant to be applicable as long as the faulty input doesn't affect the correct operation of the system.

Figure 2-11 is a familiar illustration of how someone adopting this guideline approach manages to communicate with someone who does not follow strict English grammar rules.

In order to make sense of all these design rules, we need to consider briefly how the process of new protocol design in the Internet has evolved into the procedures used by today's IETF. The following contains a brief history of the Internet for this purpose.

Figure 2-11 Be strict when sending and tolerant when receiving.

History of the Internet Protocol Development Process

The Internet's precursor, the ARPANET, grew out of some research efforts on packet-switching technology that began at the end of 1964 with just four interconnected computers.

From the beginning of the 1970s, the ARPANET used *Network Control Protocol* (NCP) for host-to-host transport. The first work on IP began during the same period, and although IP soon became the dominant network-layer protocol in the ARPANET, it wasn't until January 1983 when migration from NCP to IP finally ended. Today, the Internet is the largest IP network in the world.

Origins of the Request For Comments (RFCs)

ARPANET was developed within the research and academic community, with its strong tradition of sharing results and ideas. The research community is an open environment that creates quick development. The pace at which the ARPANET evolved proved to be extremely quick. The community quickly determined that the model of academic publications, which often takes more than a year in review and production, was too stodgy to support Internet-style growth and innovation.

In order to avoid this bottleneck, new documents called *Request For Comments* (RFCs) were created in 1969 by the community. In much the same way operators issue *Request For Information* (RFIs) and *Request For Products* (RFPs) to get feedback from vendors, individuals within the Internet community issued RFCs to get feedback and ideas from others involved in similar work. This information could be distributed fast and accessed for free.

The idea was to gather all the feedback on a specific RFC in one place and to publicly identify open issues. After all the necessary discussions, when consensus was reached, a technical specification could be developed without encountering continual obstacles.

The first RFCs were written by researchers working at the same location. Later, the appearance of e-mail enabled RFC authors to be much more widely distributed. Mailing lists gained increasing importance in the standardization process and currently, it is common to find RFCs written by people who have never met.

This story has a happy ending. The ARPANET grew unchecked. All of the individuals within the community had open access to all of the protocol specifications and other technical documents. Access to these documents enabled the rapid implementation of new applications that could readily interoperate with existing and emerging applications.

Coordination Bodies

By this point, the ARPANET's pace of growth necessitated some kind of coordination. In the late 1970s, some bodies were loosely formed for this purpose: the *International Cooperation Board* (ICB), the *Internet Research Group* (IRG), and the *Internet Configuration Control Board* (ICCB).

These bodies coped with the growth for some time, but were soon overwhelmed. That's when the task forces were first created; each task force consisted of people working on a well-specified problem area, and each had a chair to oversee progress.

The *Internet Activities Board* (IAB) was formed next; its membership consisted of all the chairmen from the various task forces. One of these early task forces was the Internet Engineering Task Force (IETF). Figure 2-12 illustrates the original structure of the IAB.

In the mid 1980s, public interest in the engineering of the Internet grew dramatically. More and more people started attending IETF meetings. In response, the IETF was divided into working groups. Every working group had one or more chairs and a predefined scope. These working groups were grouped together forming areas—each area under the jurisdiction of an area director or directors. Collectively, the area directors formed the *Internet Engineering Steering Group* (IESG). The remaining task forces were collapsed into the *Internet Research Task Force* (IRTF).

The *Internet Society* (ISOC) was created in 1991 to administer the standardization process itself rather than any specific technology. In 1992, the IAB was rechristened the *Internet Architecture Board*, and we arrived at the current configuration of Internet governance (see Figure 2-13).

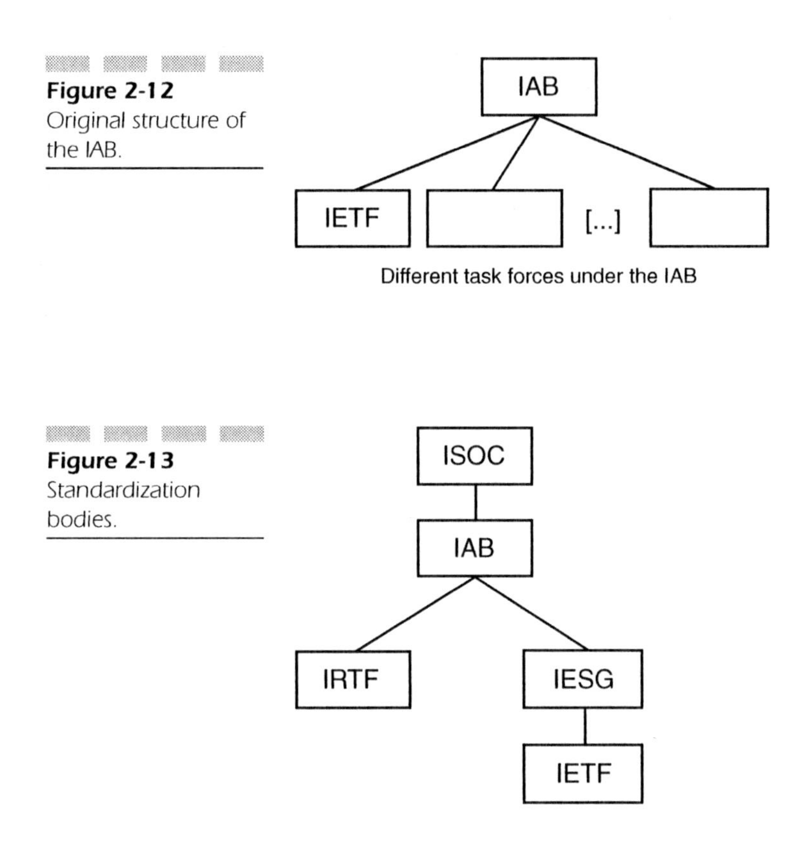

Figure 2-12 Original structure of the IAB.

Figure 2-13 Standardization bodies.

The IETF

The Internet standard process [RFC 2026] has always been evolving. Today, Internet standards are developed in the IETF—an open community of people concerned with protecting the performance of the Internet and related technologies. The IETF is divided into working groups (currently, more than 125 working groups) wherein the actual technical work is conducted.

IETF working groups are usually short lived; conversely, research groups in the IRTF are usually long lived. A working group focuses on a technology area that is defined when the group is created; it dissolves when the group has answered the specific questions posed by its charter to the satisfaction of the IESG (or when stalemated). Examples of working groups that we'll consult in this book are *Multiparty Multimedia Session Control* (MMUSIC), *IP Telephony* (IPTEL), and Session Initiation Protocol (SIP).

Working groups fall within the following nine different areas in the IETF. The IETF structure is illustrated in Figure 2-14.

- Applications
- General
- Internet
- Operations and management
- Routing
- Security

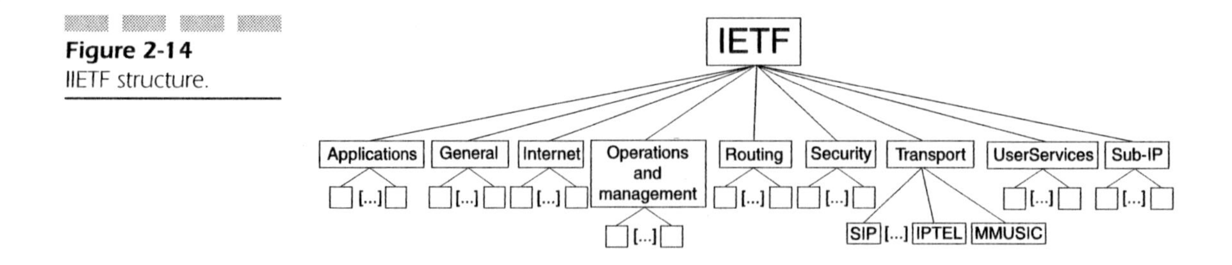

Figure 2-14
IIETF structure.

- Transport
- User services
- Sub-IP

The IESG

The IESG, composed of area directors and the IETF chair, reviews the documents produced in working groups and approves them, where warranted, as standards. In other words, the IESG is the technical management of the IETF.

The IESG also charters new working groups. If an area director considers that the IETF has enough interest and willing bodies to work on the issues proposed, a charter is written to specify working group deliverables and the tasks it's empowered to work on. A charter also includes the time frame for the outcome of the working group.

Issues can be raised at the IETF outside of the working group structure. When it's unclear whether a new problem set is or isn't a threat to the Internet or whether it warrants a working group, interested parties can hold a *Birds Of Feathers* (BOF) meeting. Area directors use BOFs to investigate the issues at hand and assess interest within the community.

The Technical Work

Each working group has one or more chairs to manage progress and procedures. Each working group also creates an official mailing list for public discussion of issues associated with the working group. Access to the mailing list is free and unrestricted, and archives are maintained as a matter of record. By now, most of the work done in a working group is carried out in the mailing list. Technical discussions about unresolved issues are conducted, agreements on procedures are reached, and general questions about the technology being developed are posed. Working groups make decisions based on rough consensus. When individuals or groups of individuals differ on a certain matter, the dominant view prevails. (How to measure dominance is an ongoing debate for the IETF.) The viewpoint prevails that a decision is more important than a resolution of all differences.

IETF Specifications: RFCs and I-Ds

The IETF publishes its documents in the RFC series. However, not all of the documents produced by IETF define standards. The RFC series has basically three types of specifications: Internet standards track specifications, non-standards track specifications, and *Best Current Practice* (BCP) RFCs. Figure 2-15 shows the three types of specifications in the RFC series. The following sections examine each type.

Standards Track RFCs RFCs following the standards track evolve through three maturity levels: proposed standard, draft standard, and standard. Specifications coming out of a working group are classified as proposed standard when they are believed to be stable and well understood. Before reaching this level, proposed standards must be reviewed carefully by the community of interest. Although some implementation experience is highly desirable, it is not formally necessary for a specification to reach proposed standard status.

This is not so for draft standards. In order to achieve the next maturity level, an RFC must point to at least two interoperable implementations in the real world. These implementations must contain all of the features that the specification describes so that unproductive features can be removed from the spec. A time requirement is also in place to ensure adequate discussion and review. Before becoming a draft standard, an RFC must stay at the proposed standard level for at least 6 months.

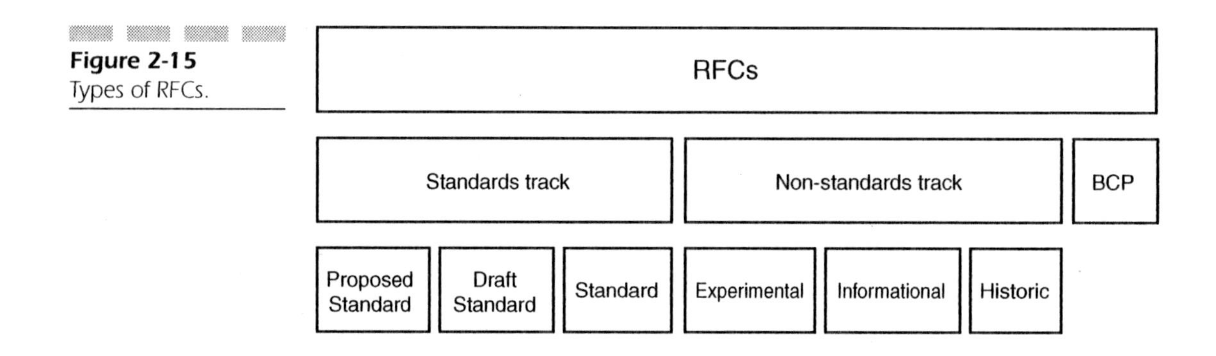

Figure 2-15
Types of RFCs.

A specification on the standards track must eventually meet two main requirements: rough consensus and running code. These two requirements are key aspects of the IETF process.

The last maturity level on the standards track is the Internet *standard* (STD) level. A specification becomes a standard only when wide implementation and operational experience appears within the community. Internet standards are stored in the STD subseries where new standards are assigned an STD number.

Every IETF specification has an associated RFC number, and when a certain specification enters the next maturity level on its track, it is immediately assigned a new RFC number. For example, the proposed standard "An Extension to HTTP: Digest Access Authentication" was originally RFC 2069. When it reached the draft standard level, it became RFC 2617 "HTTP Authentication: Basic and Digest Access Authentication." The RFC archives document and preserve the full development cycle of each protocol in this way.

When a specification reaches the Internet standard level and is included in the STD subseries, it keeps an RFC number nonetheless. For instance, the IP protocol is defined in RFC 791, but because the status of this protocol specification is Internet standard, it also appears in the STD subseries: STD0005. A specification cannot reach the Internet standard level without being a draft standard for at least 4 months, and at least one IETF meeting has to be held before a draft RFC can become an Internet standard.

Non-standards Track and BCP RFCs As suggested earlier, some IETF specifications do not define any standard at all. These specifications follow the non-standards track. Experimental RFCs show results and conclusions from research and development work. They record the experiences of implementers and designers working on some area of the technology.

Informational RFCs are used to spread general information to the community. Neither experimental nor informational RFCs are reviewed as carefully as standards track documents because they do not incorporate a proposal. All RFCs become historic when they are obsoleted by a new RFC, or when they are no longer applicable.

BCP RFCs standardize practices and give guidelines for policies and operations. BCP RFCs do not have maturity levels; their process is similar to the one for proposed standards. BCP RFCs are stored in the BCP subseries, but retain their RFC number. For instance, RFC 2026 "The Internet Standards Process" is also BCP0009.

Internet Drafts (I-Ds) Internet drafts (as opposed to RFCs) are draft documents used inside the working groups to provoke and gather feedback. They are issued before a more definitive version of the specification reaches RFC status. An Internet draft is valid for at least 6 months or until a newer version of the draft appears. Internet drafts do not define standard documents, and the IETF discourages referencing them in any way but as work in progress. Note that this books references some Internet drafts. The reader should be aware of the limited and unofficial status of an Internet draft.

Internet drafts are also employed in the standards track. When a proposed standard RFC is issued, it's accompanied by a new Internet draft. All changes and additions required for the proposed standard RFC to reach the draft standard level are executed on this new draft, which in final form becomes the draft standard RFC. The basic specification of the SIP protocol [RFC 2543] is currently a proposed standard, having reached this maturity level in February 1999. Since then, an Internet draft displaying all the modifications made to the specification since March 1999 has been posted. This Internet draft [draft-ietf-sip-rfc2543bis] will be the future SIP draft standard RFC.

Figure 2-16 shows the life cycle of an IETF specifiction—from the first Internet draft until the specification is issued as an Internet standard. Several Internet drafts with different version numbers are released until the spec reaches a particular maturity level. At that point, the specification is published as an RFC, and a new Internet draft is issued to gather feedback in order to reach the next maturity level.

It is worthwhile mentioning that some specifications, after having reached the proposed standard maturity level and having gathered feedback in a new Internet draft, are re-issued as a new proposed standard RFC rather than as a draft standard RFC. This happens when the specification has changed substantially from the last RFC that was issued so a new RFC is needed but the new RFC cannot have draft standard status since the specification is not mature enough.

Naming Internet Drafts As described previously, the charter of a working group contains a limited set of deliverables that the working group agrees to work on. With these deliverables in mind, the group defines a set of working group *items*. When the working group releases a draft about a particular item, it is named according to the following format: draft-ietf- followed by the name of the working group followed by version number.

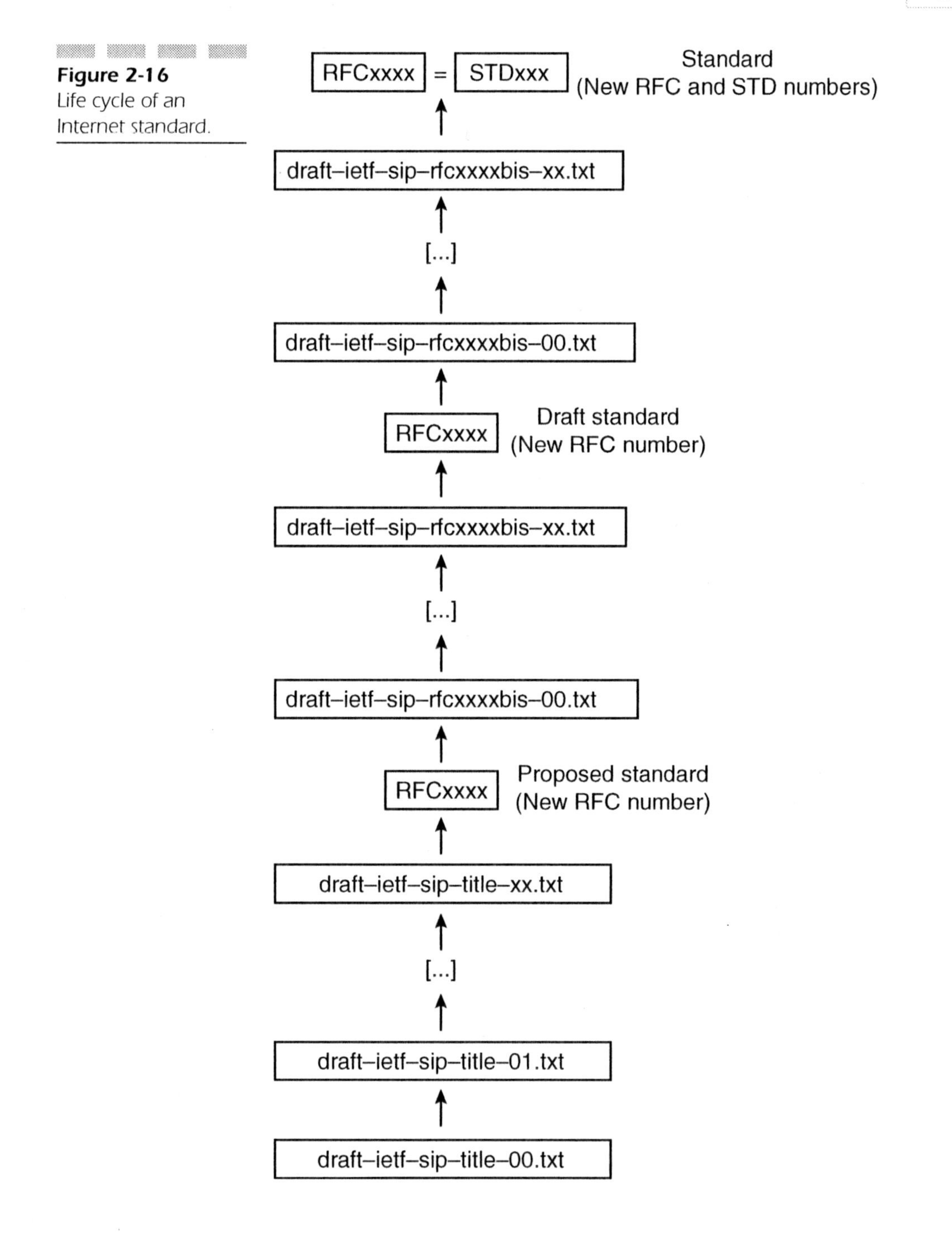

Figure 2-16 Life cycle of an Internet standard.

Naming conventions term the first release of an Internet draft version 00. Hence, draft-ietf-sip-call-flows-01.txt is the second version (version number 1) of this specific Internet draft belonging to the SIP working group.

Individual contributions to a working group discussion that are not directly related to a working group item or that are preliminary versions of a draft on a working group item follow a slightly different naming convention containing the author's name: for example, draft-schulzrinne-sip-911-01.txt.

Finally, Internet drafts and RFCs can be found in different formats (PostScript, *Portable Document Format* [PDF], txt), but the definitive reference is always the ASCII text version. It includes all the figures, tables, illustrations, and diagrams contained in the specification.

Figure 2-16 outlines the life cycle of an Internet standard. A new idea (e.g., a new protocol) is first documented in draft-ietf-sip-title-00.txt. This very first draft evolves thanks to the feedback given by the community until it reaches the standard (STD) maturity level.

However, not all the drafts reach the end of the standardization process shown in figure 2-16. Many drafts reach the proposed status, but just a few proposed RFCs become a draft standard later. Among those that become a draft standard, even fewer become Internet standards. It is common for a proposed RFC to gather feedback and to be re-issued as a proposed RFC again (with a new RFC number) rather than as a draft standard.

In the following chapters we will study different protocols and extensions to protocols specified by the IETF. It is important to keep Figure 2-16 in mind in order to understand in which maturity stage a particular specification is at present.

The Internet Multimedia Conferencing Architecture

Now that we have stipulated how the *Internet Engineering Task Force* (IETF) works and some general characteristics found in all the protocols designed by the IETF community, *Session Initiation Protocol* (SIP) among them, we can turn our attention to the context in which SIP is used. We will first see which protocols are used below SIP in the transport layer and which functionality they provide.

We'll also discuss what motivated the *Internet Engineering Steering Group* (IESG) to charter the SIP working group in the first place, in order to understand where SIP is intended to work. We'll focus on the Internet multimedia conferencing architecture. This architecture includes a set of Internet protocols that together provide multimedia services in the Internet environment. SIP is part of this architecture, which proves to be one of SIP's strengths. As such, it interacts smoothly with the rest of the protocols and exploits their functionality; that is, SIP is one tool inside the toolkit that this architecture represents.

The Internet Layered Architecture

We have said that one of the most important strengths of the Internet is its suitability to be used as a service creation platform. The IETF adheres to a layered approach to create services. The Internet layered model consists of four layers that are implemented on top of the physical layer (Figure 3-1), which are commonly known as the *Transmission Control Protocol / Internet Protocol* (TCP/IP) protocol suite.

A transport layer is implemented on top of the common IP layer and application layer protocols make use of the different transport protocols. A particular application aiming to provide a certain service picks up the application layer protocols required to implement the service.

Let's say we want to build a Web browser that enables users to surf the Internet and to read e-mail. To do so, we will have to pick up two application layer protocols. We choose *Hypertext Transfer Protocol* (HTTP) [RFC 2068] to surf the Internet and *Internet Message Access Protocol* (IMAP) [RFC 2060] for downloading e-mail from a server. Both run on top of the Transmission Control Protocol (TCP) [RFC 793], which is, of course, a transport layer protocol (Figure 3-2).

Figure 3-1
Four layers of the TCP/IP protocol suite.

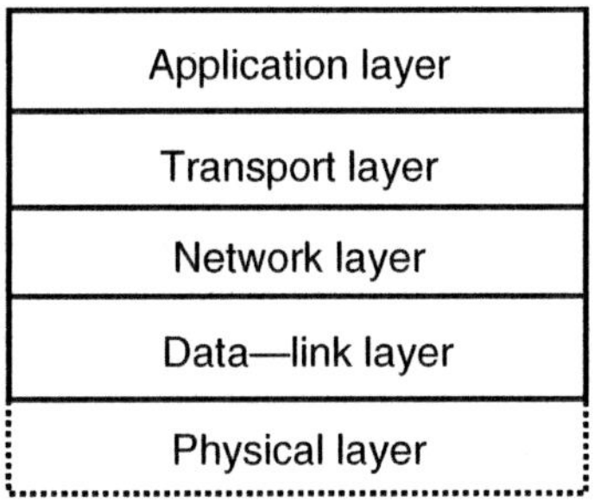

Application layer
Transport layer
Network layer
Data—link layer
Physical layer

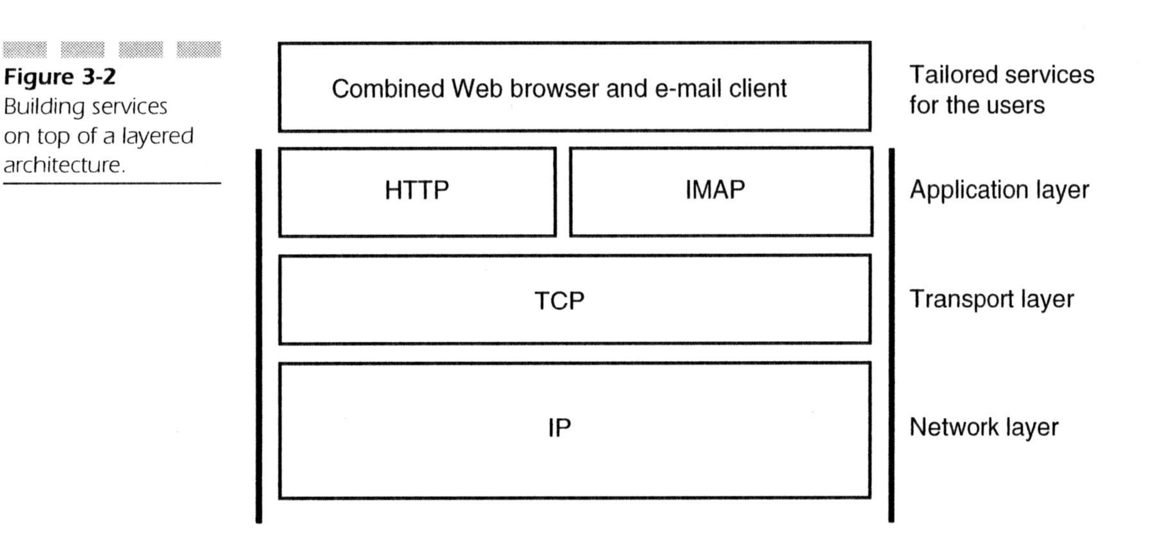

Figure 3-2
Building services on top of a layered architecture.

Transport Layer Protocols

Applications must make use of application layer protocols in order to provide any service whatsoever. Application layer protocols must in turn make use of transport layer protocols, and transport layer protocols run on top of

IP. Currently, two transport layer protocols are widely available: TCP [RFC 793] and *User Datagram Protocol* (UDP) [RFC 768].

Transmission Control Protocol (TCP) TCP provides reliable, in-sequence transport of byte-streams between hosts. It contains mechanisms such as time-outs, retransmissions, and sequence numbers that let the protocol deliver data to the receiver exactly as generated by the sender. TCP also performs flow control and error correction.

Thus, if Bob sends Laura the message "Laura, how are you doing?" over a TCP connection, TCP will ensure that Laura receives the message "Laura, how are you doing?" Because TCP is an end-to-end protocol, it deduces the state of the network by observing the dynamic behavior of the end-to-end traffic. Receivers send acknowledgement messages to senders to provide enough information so that error detection and flow control can be performed.

TCP enables the end user to demultiplex incoming IP packets to different applications. Because all of the packets arriving at a host contain the same destination IP address, a further identifier is needed to associate each arriving packet to the proper application. These identifiers are referred to as TCP port numbers, and each IP-based application uses one. For instance, datagrams with the destination TCP port number 80 are delivered to the HTTP application, whereas datagrams sent to the TCP port number 23 are handled by the Telnet [RFC 854] application (Figure 3-3).

User Datagram Protocol (UDP) UDP [RFC 768] provides unreliable datagram delivery; that is, it does not ensure that a given datagram will arrive at its destination by any means. UDP merely performs IP traffic demultiplexing based on UDP port numbers. Both UDP and TCP are 16-bit port numbers. After traffic demultiplexing, UDP provides a checksum that allows end systems to check that the datagrams received were not corrupted by the network.

For instance, if Bob sends the same message as before using UDP, he cannot be sure Laura will receive it. To find out, he'll have to check with Laura herself. If a datagram containing a UDP packet gets lost, Bob will also have to retransmit it himself; UDP does not provide such functionality.

Stream Control Transmssion Protocol (SCTP) The *Stream Control Transmission Protocol* (SCTP) [RFC 2960] is a newly developed transport protocol. It is foreseen that SCTP will be widely implemented, but currently, it is less available than TCP and UDP.

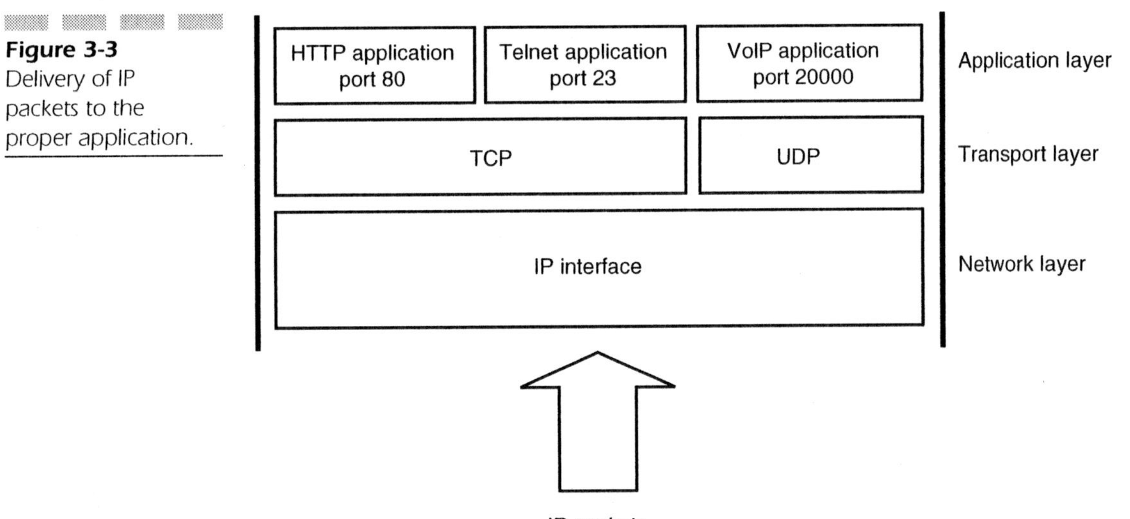

Figure 3-3 Delivery of IP packets to the proper application.

Real-Time Services in the Internet

User services are implemented using application layer protocols. Two of the most familiar Internet services are the Web and e-mail. The Web uses the application layer protocol HTTP and e-mail is implemented using, among other protocols, *Simple Mail Transport Protocol* (SMTP) [RFC 821] and Internet Message Access Protocol (IMAP) [RFC 2060].

These services, and others like them, have boosted the Internet development for several years. They consist of an asynchronous exchange of information. The Internet has been proved to be a powerful tool in developing such asynchronous services, but it can also provide synchronous (real-time) services. Examples include videoconferencing over the Net and live broadcast on a workstation attached to an IP network. Real-time services are delay sensitive. The information they carry needs to be delivered to its destination within a prescribed time limit. If the delay introduced by the network is any longer, either the information becomes no longer useful to the receiver or the service quality drops dramatically.

It is possible to further classify real-time services into streaming and interactive categories (see Figure 3-4). Streaming services typically have

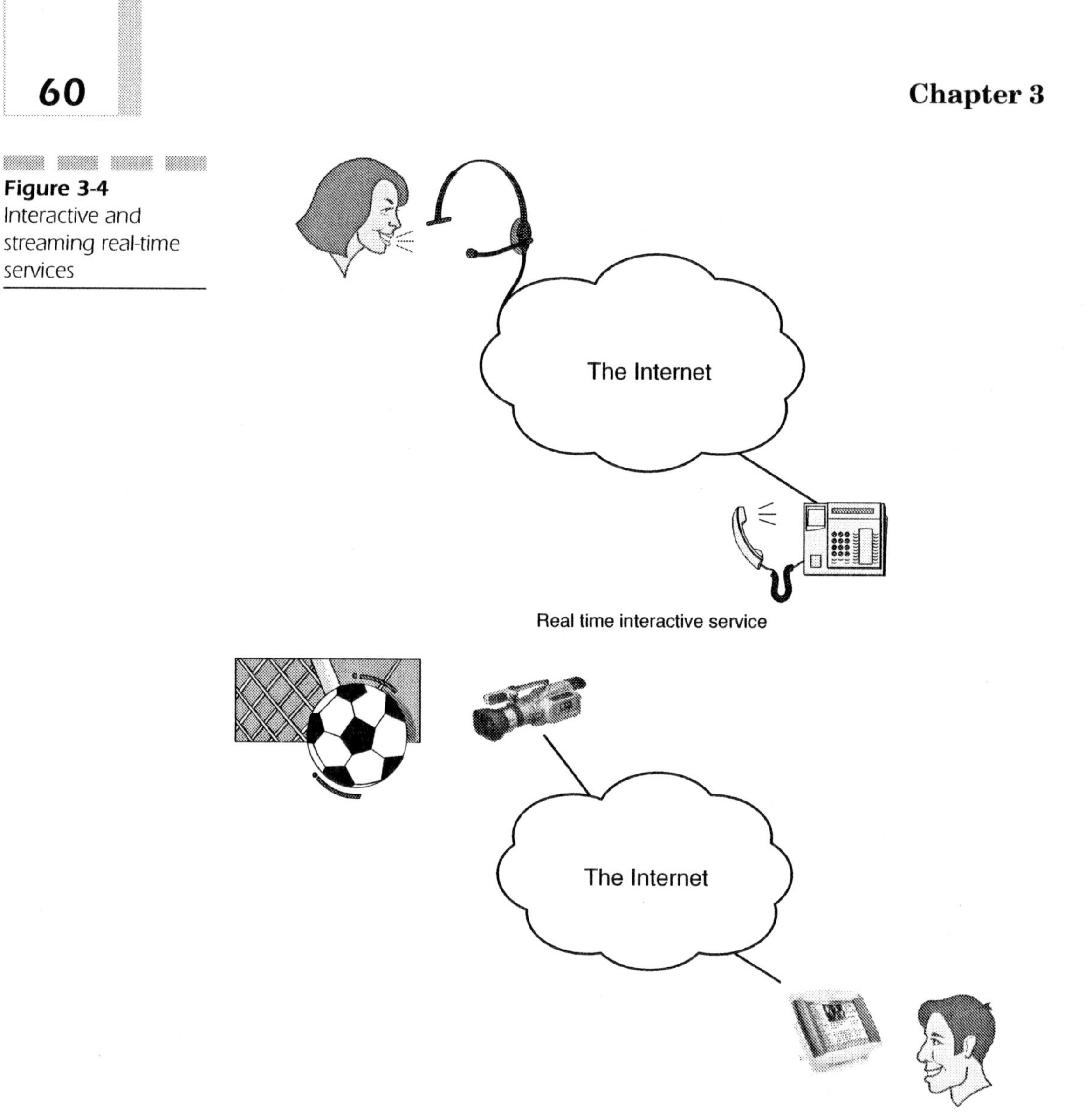

Figure 3-4 Interactive and streaming real-time services

lower requirements than interactive ones. Consider the transmission of a soccer match as an example of a streaming service. Relatively large delays, in the order of even some seconds, are acceptable for the user as long as the image and sound quality hold up, but probably not vice versa. The industry currently deems it acceptable for users to see a goal on the screen a couple of seconds after it really happened on the soccer field where the match takes place.

Services with an interactive flavor have tougher requirements. In a voice conversation over the Internet, the delay must be kept very low. Otherwise it is impossible to undertake a normal conversation. The maximum acceptable delay for this kind of real-time interactive service depends on many factors, but in general, it is lower than the delay for streaming services. According to the *International Telecommunication Union Telecommunication Standardization Sector* (ITU-T), the maximum acceptable delay for voice conversations is a round-trip delay of 300 ms.

The Internet Multimedia Conferencing Architecture Advanced real-time services include several types of media (for example, videoconferencing includes video streams and audio streams) and are therefore referred to collectively as *multimedia services*. The IETF has developed a set of protocols specific to multimedia services. Applications will have to combine some of these protocols in order to attain the required functionality. Figure 3-5, taken from [draft-ietf-mmusic-confarch], shows how all these protocols fit together. SIP is part of the Internet multimedia conferencing architecture and is shown in real context in Figure 3-5. If you take the time to understand the role of each protocol in the architecture, you'll have a good functional explanation of why SIP was needed and what it is expected to deliver.

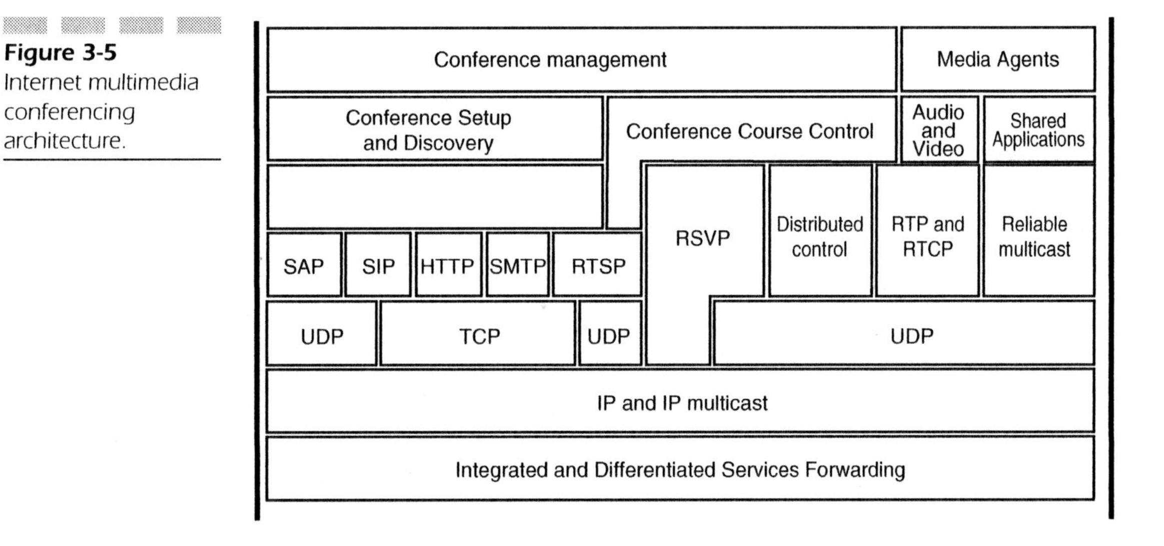

Figure 3-5 Internet multimedia conferencing architecture.

Multicast

The Internet is designed for best-effort delivery of datagrams between hosts. A datagram contains a destination IP address and the routers in the network undertake the task of moving the datagram towards its destination. An IP address typically identifies a network interface in a host (such as an Ethernet card). This mechanism enables two-party data exchange between any two hosts in the network, so long as one host acts as sender and the other as receiver. It is also possible for the receiver to return data to the sender just by noting the sender's IP address. The kind of routing where one IP address identifies a single-host interface (Figure 3-6) is referred to as *unicast*.

Routing Towards Many Receivers

This simple scenario gets complicated when more than two parties wish to communicate. In order to send an IP packet to all of the hosts

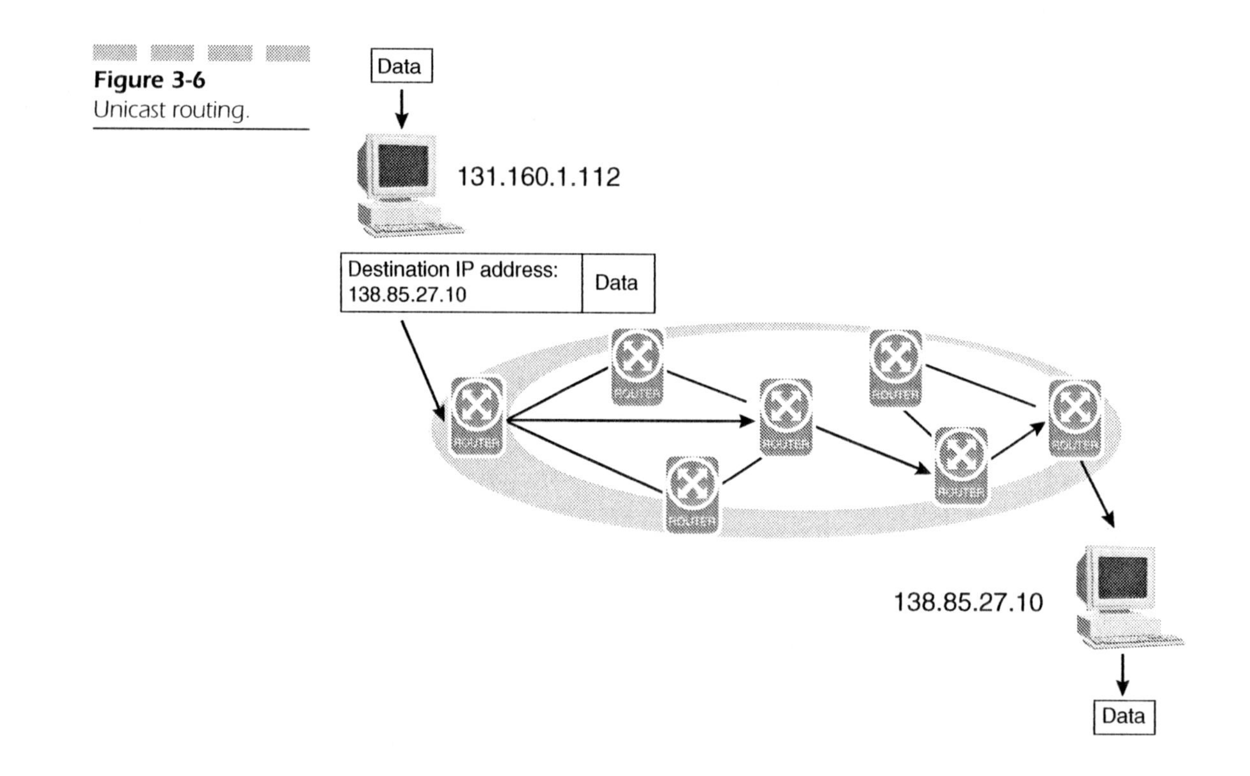

Figure 3-6
Unicast routing.

involved in a n-party conference, a host would have to transmit n-1 IP packets, each containing the IP address of one of the hosts designated to receive the data.

Obviously, systems with these mathematics don't scale well for a large number of users. As soon as the number of hosts involved increases at all, traffic in the network increases dramatically, straining the processing power of the hosts. Every host needs to register the IP addresses of all receivers and build n-1 IP packets with exactly the same information— the only difference being the destination address. Many of these packets traverse a common set of nodes until they finally are routed to their destinations, requiring links in the network to transport the same information multiple times (Figure 3-7).

In Figure 3-7, packets from 131.160.112 to 138.85.27.10 and packets from 131.160.1.112 to 153.88.251.19 follow the same path to arrive at the router R. Therefore, all the links from 131.160.112 to R are loaded twice as much as they ought to be.

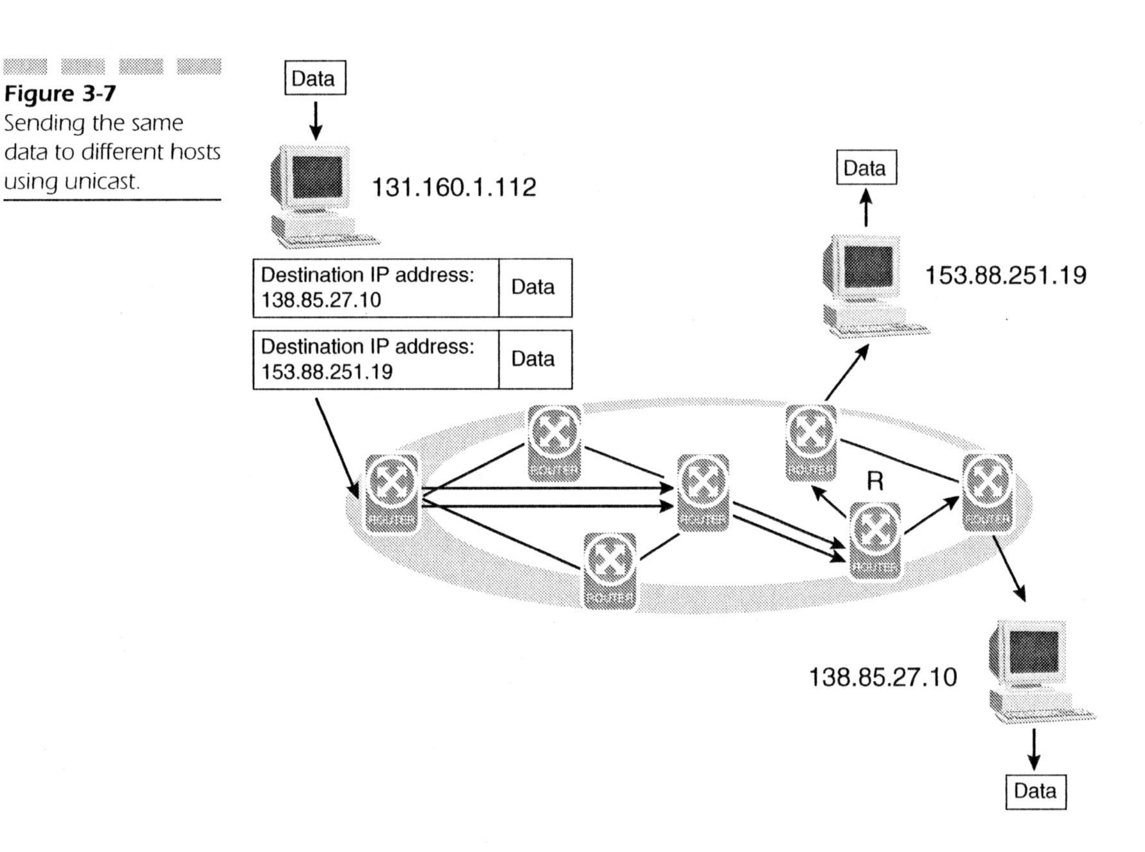

Figure 3-7 Sending the same data to different hosts using unicast.

Multicast routing resolves these issues by providing many-to-many communication scaling to groups with a large number of participants. These routers collaborate to forward IP packets destined for a multicast address towards their several destinations. A multicast address represents a group of hosts willing to receive the IP packets that will be transmitted, and by the same token, hosts receiving data from a multicast address are members of a particular multicast group. Any host in the network can send data to a multicast address, but it must belong to the multicast group in order to receive data. (In IP version 4 [RFC 791], addresses between 224.0.0.0 and 239.255.255.255 are reserved for multicast.)

Advantages of Multicast

Multicast is scalable because data traverses each link only once (Figure 3-8). Multicast routers will duplicate IP packets only if necessary and even then will duplicate as close to the members of the multicast group as possible.

Figure 3-8 shows the same configuration as Figure 3-7. Packets from 131.160.1.112 to 138.85.27.10 and from 131.160.1.112 to 153.88.251.19 are also sent, but, this time, via multicast routing. Links from 131.160.1.112 to R carry the same data just once. R will duplicate the data in order to send it to 138.85.27.10 and 153.88.251.19. This example illustrates two main savings:

- Half of the bandwidth used from 131.160.1.112 to R is saved.
- The end systems do not need to know who the members of the multicast group are in order to send data.

Senders send data to the multicast address without knowing the identity of all the receivers. Only the routers closest to the receivers know which hosts are members of the multicast group. No central server is keeping track of multicast groups; all the information is distributed among multicast routers.

The most a multicast router needs to know is whether at least one member of the multicast group can be reached using one of the router's interfaces. If the answer to that question is yes, the data will be transmitted through this interface.

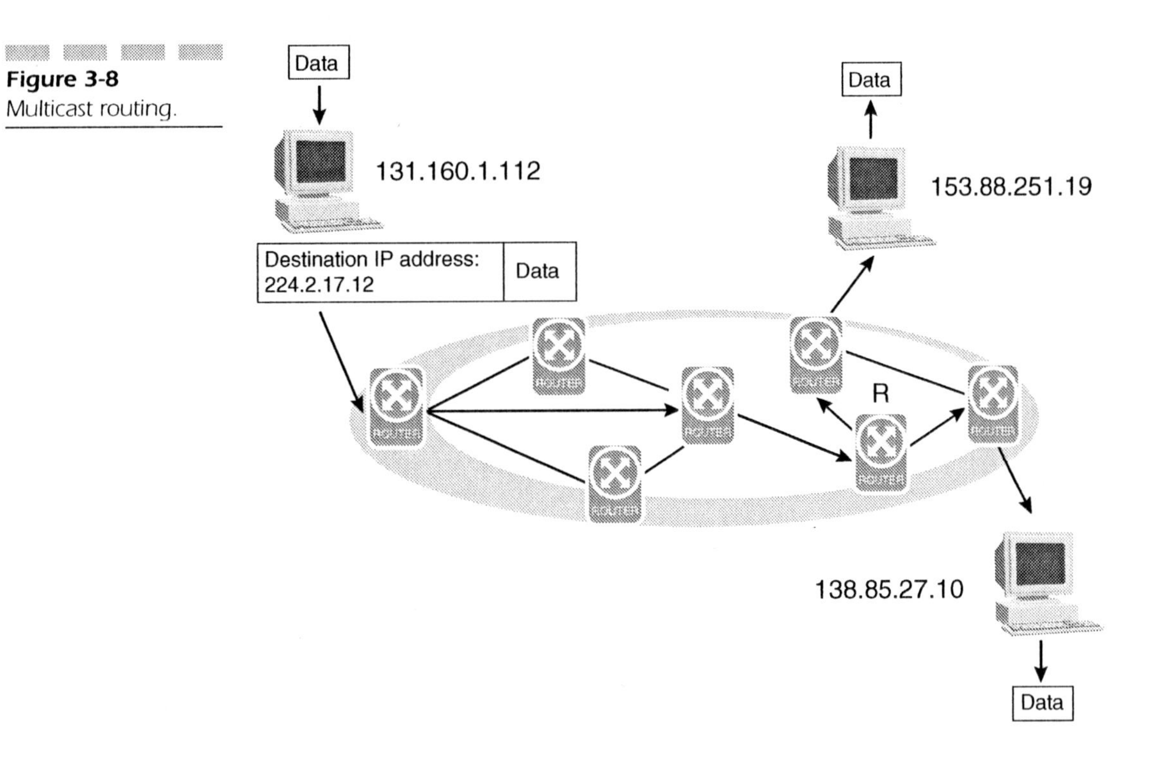

Figure 3-8 Multicast routing.

Multicast Routing Protocols

Multicast routers use multicast routing protocols to build distribution trees from senders to receivers. These protocols can be grouped into two categories, sparse mode and dense mode. These two modes use different algorithms to build trees.

Dense Mode Multicast Routing Protocols Dense mode multicast routing protocols work well in networks where most of the hosts are members of the multicast group. When the proportion of hosts receiving multicast data is high, the probability of having at least one member of the multicast group in any particular subnetwork is also high.

Dense mode protocols usually build a shortest-path tree per sender or group of senders. Thus, a different distribution tree is used depending on where the data comes from, which is why they're usually referred to as source-based trees. Figures 3-9, 3-10, and 3-11 show how three different distribution trees are calculated for three different senders.

Some common examples of dense mode protocols are *Distance Vector Multicast Routing* (DVMRP) [RFC 1075] and *Protocol Independent Multicast-Dense Mode* (PIM-DM). DVMRP is currently enjoying the most widespread use.

Sparse Mode Multicast Routing Protocols Sparse mode protocols are better suited for networks where the proportion of hosts receiving multicast traffic is low. These protocols usually implement a rendezvous point and, based on it, build a shared tree that will be used by all sources.

Figure 3-12 shows how a distribution tree for the rendezvous point is built. The tree is computed as if the rendezvous point were the sender and all the other group members were the receivers. While this tree is being computed, packets from the group's senders are encapsulated and routed towards the rendezvous point, which then delivers them to receivers by consulting the tree.

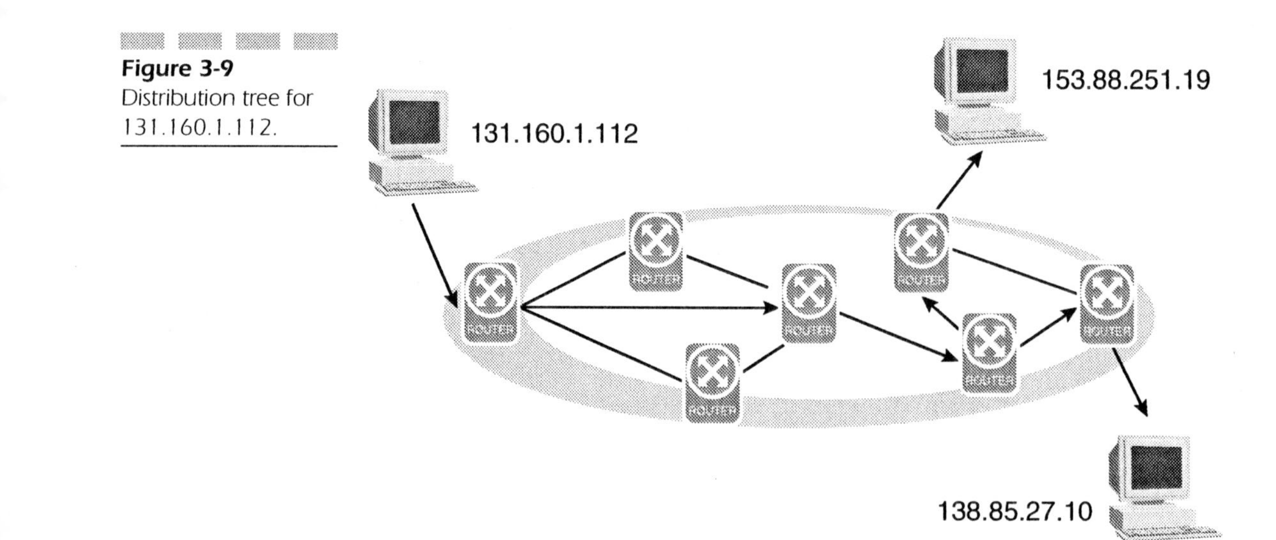

Figure 3-9
Distribution tree for 131.160.1.112.

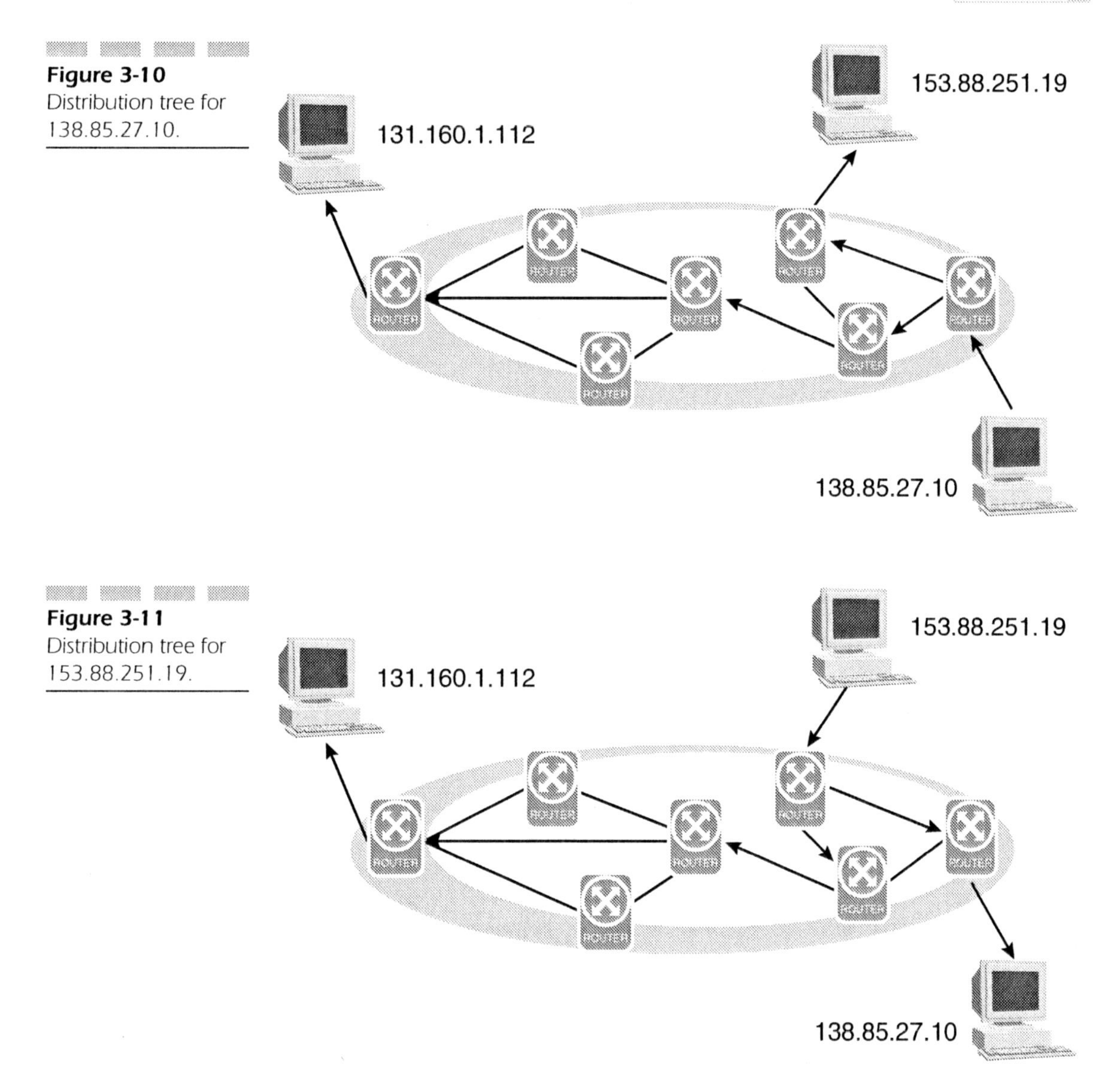

Figure 3-10 Distribution tree for 138.85.27.10.

Figure 3-11 Distribution tree for 153.88.251.19.

Once a tree for the rendezvous point is complete, all nodes use it. This means that packets are no longer encapsulated for transmission to the rendezvous point, but are routed towards receivers directly through the tree. We can see in Figure 3-13 that the shared distribution tree used by nodes

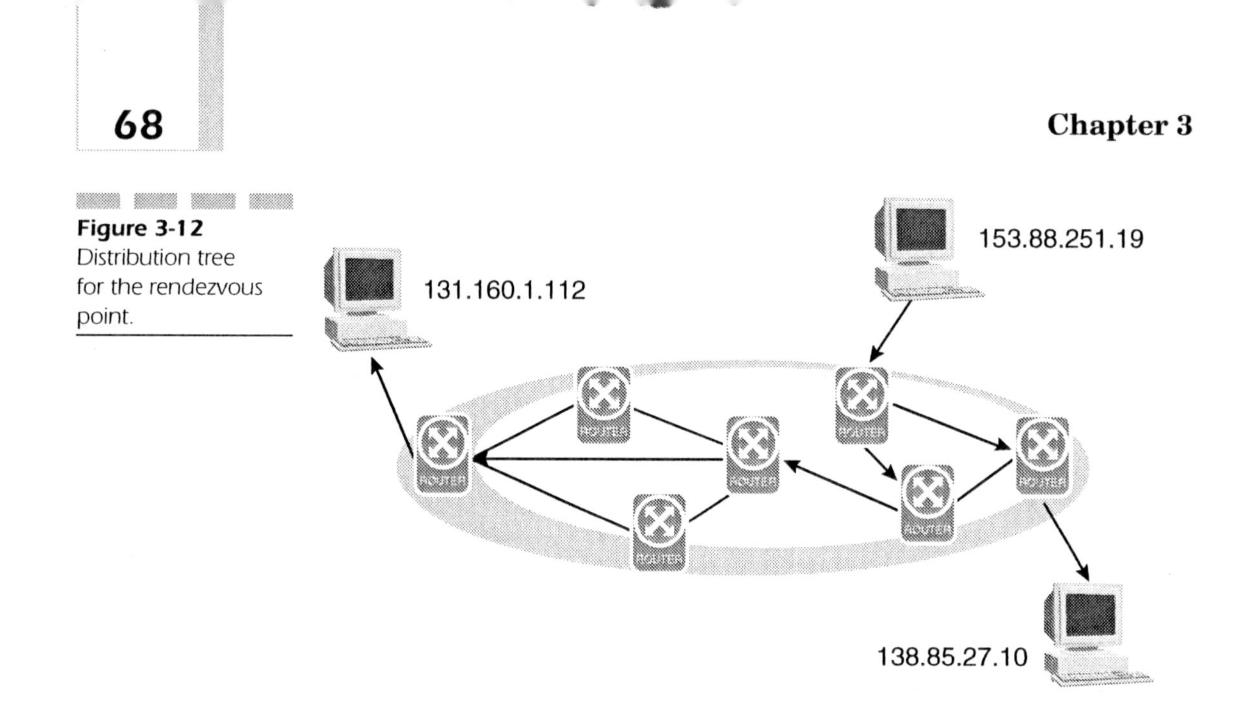

Figure 3-12 Distribution tree for the rendezvous point.

with IP addresses 138.85.27.10 and 153.88.251.19 differs from the one calculated for them specifically in Figure 3-13.

Examples of sparse mode protocols are *Protocol Independent Multicast-Sparse Mode* (PIM-SM) [RFC 2362] and *Core-Based Trees* (CBTs) [RFC 2189].

IGMP

Apart from the protocol used to build the distribution tree, a host wishing to receive multicast data needs to become a member of a particular multicast group —a set of hosts that subscribe to the same multicast address. *Internet Group Management Protocol* (IGMP) [RFC 2236] [draft-ietf-idmr-igmp-v3] is used for this purpose (see Figure 3-14). Hosts send requests for joining or leaving a particular group. With this membership information in hand, a multicast router handling a certain subnetwork knows whether it needs to receive multicast data or not. If its subnetwork has no members, it does not need to receive datagrams addressed to the multicast address of the group and can accordingly remove itself from the distribution tree using a multicast routing protocol. Later, if a host in its subnetwork wishes to become a member of the multicast group, that router will be added to the distribution tree again.

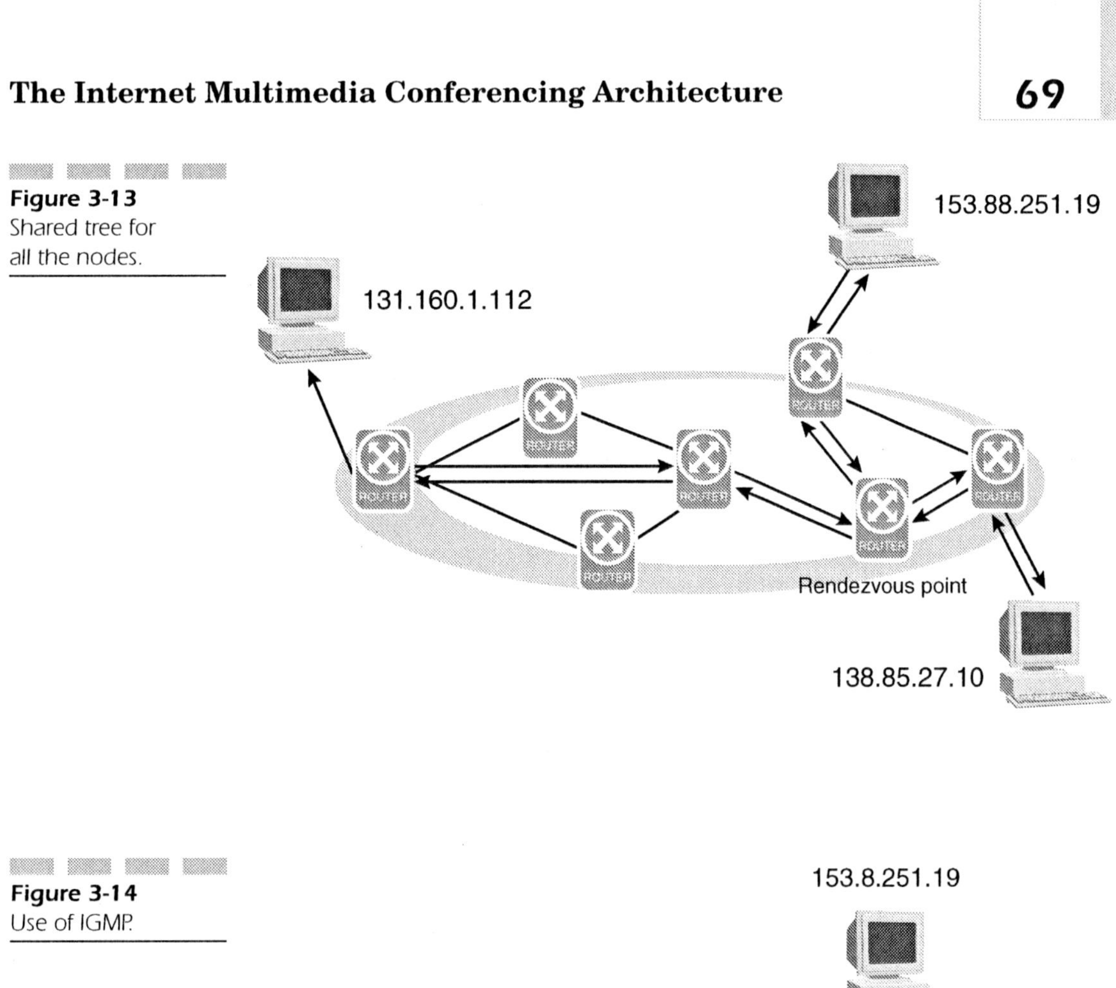

Figure 3-13
Shared tree for all the nodes.

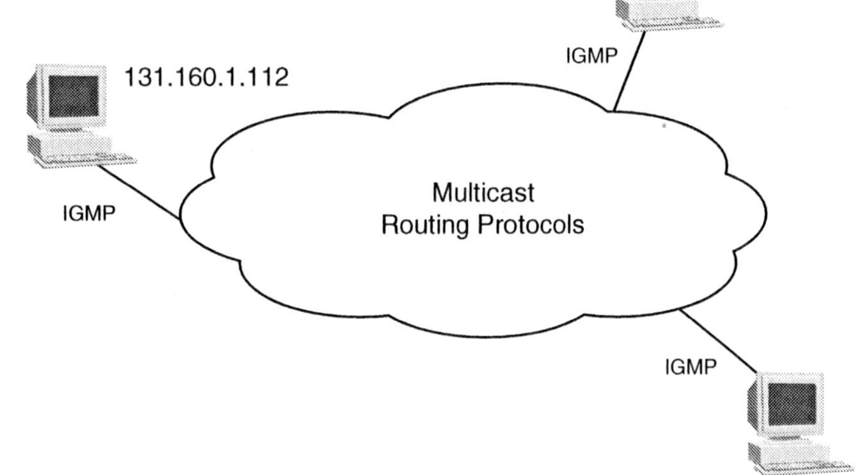

Figure 3-14
Use of IGMP.

The Mbone

The portion of the Internet that supports multicast is called the Mbone. The Mbone is *not* a separate network from the Internet. Mbone routers are also part of the Internet and are enabled for both unicast and multicast traffic. When an island of multicast routers that are not directly connected to the Mbone forms, IP tunnels are implemented between them and the rest of the Mbone. We will see that SIP was first designed for use in the Mbone to invite users to multicast sessions.

IETF sponsors ongoing work on interdomain multicast routing protocols such as *Border Gateway Multicast Protocol* (BGMP) [draft-ietf-bgmp-spec], anticipating that they will provide a more scalable multicast routing infrastructure that introduces hierarchy. Traffic aggregation between domains will reduce the state information kept by multicast routers.

Transport of Real-Time Data: RTP

The system requirements of different types of traffic are often very different. This disparity necessitates the implementation of different protocols to transport data with different requirements. For data types that need reliable transport across the network — an e-mail, for example — the sender wants the recipient to read exactly what he or she wrote, but is not equally invested in preventing delivery delay. For most e-mail, the difference between 10 seconds and 1 minute is not significant. A delayed correct packet is acceptable in this situation, whereas an on-time incorrect packet would not be.

As previously noted, the requirements for real-time traffic look quite different. Let's say that two parties embark on a voice conversation through the Internet. The encoded audio will have to be transmitted in IP datagrams to be played back to the recipient. Using common sense, a delayed correct packet in this scenario is worse than unacceptable; it's literally useless. If an out-of-sequence datagram containing what the other party said five seconds ago arrives, it will be discarded.

Jitter and Sequencing of Datagrams

Beyond these general real-time requirements are some lesser issues that a protocol designed for real-time traffic must address: jitter and sequencing of datagrams. IP networks introduce some delay to every packet traversing them. The amount of delay depends on many factors, one being the state of every particular router at the moment of packet reception. If a router has a heavy load, the packet will wait in a queue; if the queues empty, the packet will be routed immediately.

Because router state is never the same for every packet belonging to the same flow, we have a term for the variation in delay: *jitter*. If the jitter is high enough, a packet launched later than another can arrive before it, creating the out-of-sequence circumstance anathema to real-time transmissions (Figure 3-15).

To reduce jitter, the community widely employs the *Real-time Transport Protocol* (RTP) [RFC 1889]. RTP counteracts the effects of jitter and the consequent arrival of out-of-sequence datagrams by assigning timestamps and sequence numbers to the packet header. A sequence number in the RTP header enables the receiver to order the RTP packets received. Once ordered, the original timing relationship of the data contained in the payload (such as audio or video) can be recovered by reading timestamps (Figure 3-16). In the case of encoded audio, timestamps inform the receiver when to play the payload of the RTP packet through the speaker. A field called *payload type* describes which kind of data is transported in the RTP packet (such as audio encoded using a PCM codec).

Besides bearing information about payload, RTP headers also contain identifiers for the source originating the payload. Let's revisit the example of voice conversation and unpack all the operations involved. The sender produces RTP packets with encoded audio as the payload. The receiver implements a buffer to store incoming RTP packets. They are ordered according to sequence number. An RTP packet is removed from the buffer when its timestamp indicates that it is time to play back its payload.

If the timestamp of an incoming RTP packet indicates that its payload should have already been already played, the packet is discarded. Therefore, buffers should be long enough so that packets have time to arrive

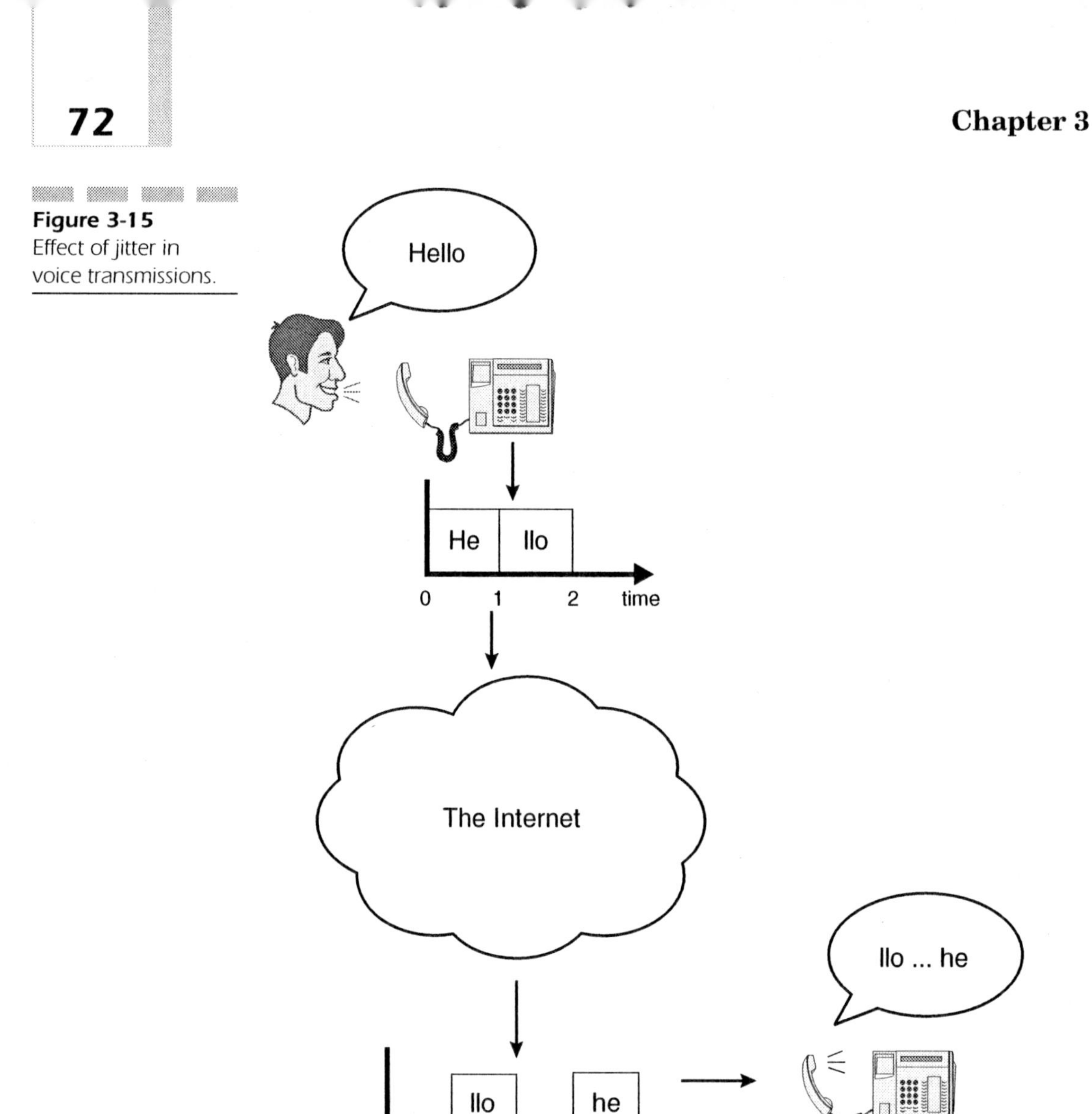

Figure 3-15 Effect of jitter in voice transmissions.

before they have to be played. Additionally, buffers should be short enough so that the inevitable delay doesn't make normal conversation impossible.

What if the packet has not arrived by the time it should be played? Some other sound is played instead. Interpolation techniques are available to create a smooth sound transition that is able to hide these gaps from the recipient.

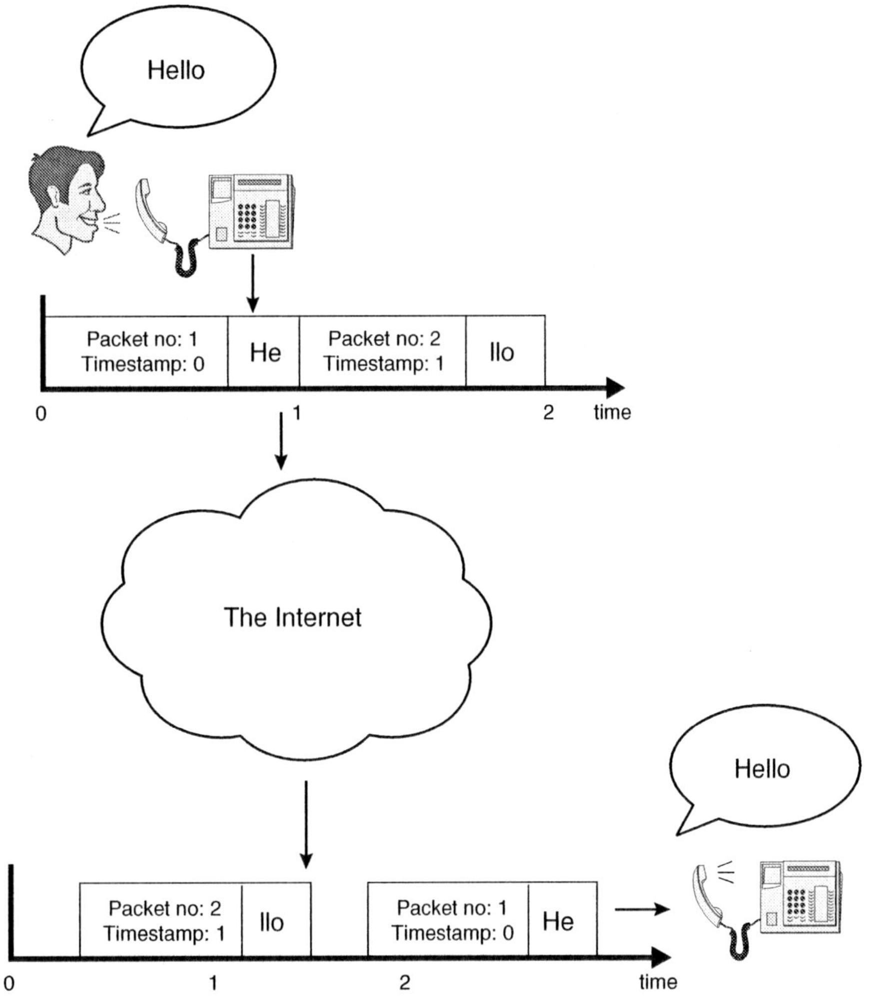

Figure 3-16
RTP couteracts the effects of jitter and the arrival of out-of-sequence packets.

Real-Time Transport Control Protocol

The previous example had just one media stream: an audio stream. Timestamps can also be used to recover the original timing relationship of the data in multiple media streams; they perform this task for each media stream individually. For instance, in a videoconference, the timestamps of the video stream ensure that the video is not played faster or slower than the original, while the audio timestamps do precisely the same with the audio stream. However, a mechanism is needed to synchronize streams, and that's where the *Real-time Transport Control Protocol* (RTCP) [RFC 1889] comes in. Its function is to associate timestamps and a real-time clock.

Every RTP session has a parallel RTCP session. Besides media synchronization, RTCP provides information about the members of the session and the quality of the communication. RTCP reports how many packets the network dropped during the session so the sender knows what quality of reception the receiver is experiencing.

QoS Provisioning: Integrated Services and Differentiated Services

The Internet's best-effort service model works fine for most applications when the network is under a reasonable load. However, when an IP network experiences heavier loads, best-effort service may not be adequate for end-to-end traffic. As delays increase, the network becomes lossier. IP packets bombard the routers faster than they can cope with them, forcing a queue to develop. When queue size limits are exceeded, the router responds by dropping datagrams.

For TCP traffic, dropped datagrams mean more retransmissions, leading to lower performance and poor *Quality of Service* (QoS). The user begins to observe end-to-end delay even with relatively forgiving applications such as file transfer.

If the dropped datagrams belong to real-time traffic, the receiver will never be received and the datagrams will not be retransmitted. As a result, the quality suffers. In the case of voice transmissions, audio can quickly become unintelligible on the receiver's end.

For applications needing a better-than-best-effort service, two different approaches fit the bill: integrated services [RFC 1633] and differentiated services (DiffServ) [RFC 2475].

Integrated Services

The basic idea behind the integrated services architecture is to give different treatment in the routers to packets belonging to different flows. For instance, datagrams from a real-time flow can be tapped for forwarding ahead of datagrams from a low-priority flow. Routers handling a particular flow will need additional information in order to forward its datagrams properly: namely information on how to distinguish the datagrams in one flow from datagrams belonging to other flows and what class of treatment each datagram is due.

A router might prioritize all datagrams with a certain destination IP address and a certain UDP destination port number, while continuing best efforts for datagrams with a different destination. In the latter case, a router handles several flows and therefore has to implement different packet filters in order to classify the datagrams it receives. The default is best effort, which is what packets that do not belong to any flow known by the router receive.

Services Available The integrated services architecture currently provides two services: controlled-load and guaranteed services. They represent two levels of better-than-best-effort services (Figure 3-17).

Packets benefiting from controlled-load services are given priority over best-effort traffic. Therefore, even if the network is congested with best-effort traffic, the controlled-load service delivers the datagrams as if the network were under moderate load. This service does not guarantee any particular bandwidth or delay for a certain flow; it simply ensures that packets get superior treatment.

Guaranteed service, as the name suggest, provides a certain bandwidth and a delay bound for a particular flow. Therefore, the jitter observed for guaranteed traffic is small to negligible.

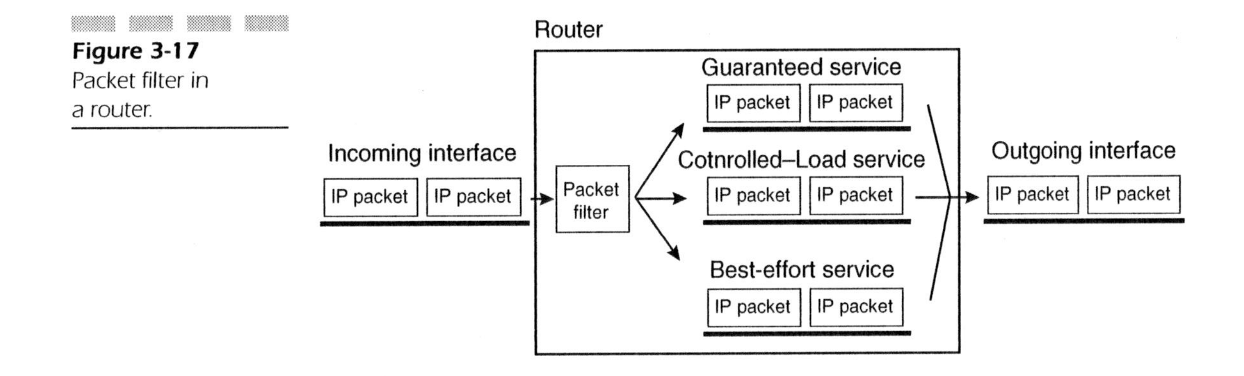

Figure 3-17 Packet filter in a router.

Because a router cannot grant high-priority treatment to an unlimited amount of traffic, routers need the capability for admission control and resource reservation.

Upon receiving a request for handling a new flow, a router checks whether it has enough available resources to accept it without impacting other flows in progress. If the QoS requested for the flow can be granted, it reserves resources for the newcomer.

State Information Stored in the Network We've just seen that routers must store information about flows in order to differentiate datagrams properly. This implies that the network stores state information. Previously, we argued that doing so flies in the face of the IP paradigm, pushing the intelligence to the end systems and storing as little state as possible in the network. The trade-off for less information is more robust systems that tolerate network failures better. Acknowledging the value of the paradigm and the need to make exceptions to it on well-defined occasions, reservation merging and soft states help minimize the problems that exceptions can cause.

Reservation Merging Figures 3-18, 3-19, and 3-20 show how reservation merging is performed in a multicast group. We saw in Figure 3-13 how the shared distribution tree was calculated for that topology. In Figure 3-18, we now assume that 131.160.1.112 is the sender and the other hosts, 153.88.251.19 and 138.85.27.10, are the receivers.

Receiver 138.85.27.10 requests a certain QoS for its incoming flow. The routers in the path store the necessary state in their packet filters and the QoS requested is honored (Figure 3-19).

Now receiver 153.88.251.19 also requests QoS for the flow it is receiving. However, the second host won't need to request QoS for the entire path from the sender because a QoS reservation is already present in part of the path for the same flow. Therefore, when 153.88.251.19 requests QoS, state information stored in the routers in the path that was already provisioning QoS for the first host does not increase. New state information is limited to those routers in the path from 153.88.251.19 to the main distribution tree (Figure 3-20).

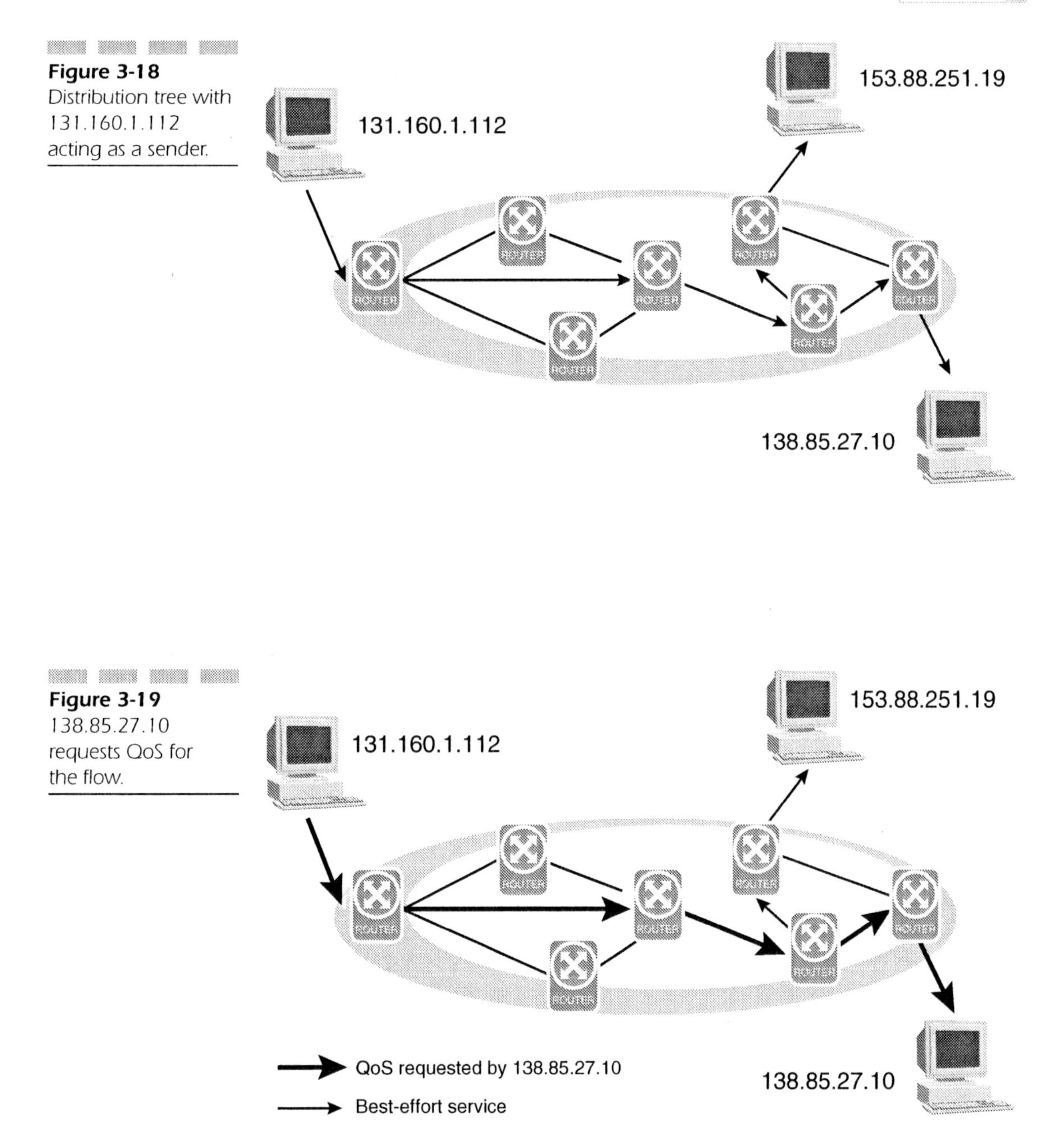

Figure 3-18 Distribution tree with 131.160.1.112 acting as a sender.

Figure 3-19 138.85.27.10 requests QoS for the flow.

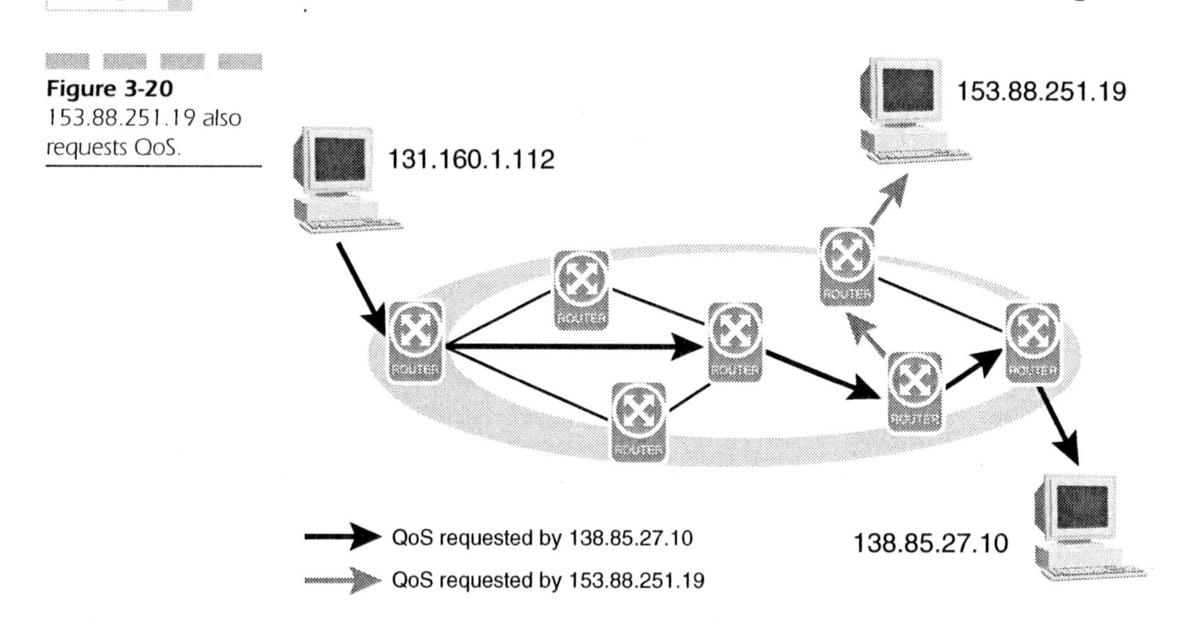

Figure 3-20 153.88.251.19 also requests QoS.

Soft States Implementing soft states increases the system robustness. Soft states store state information temporarily, after which the router removes all the state it was storing. Under this system, if the state information is not refreshed periodically, it will expire and enable the release of all the resources reserved in the router. Soft states are typically refreshed by sending a message to the router with the proper information, whereas hard states are stored permanently and require a release-of-state command to relinquish resources.

ReSerVation Protocol (RSVP) *ReSerVation Protocol* (RSVP) [RFC 2205] is the protocol used for resource reservation in the network. RSVP installs the necessary state in the routers and refreshes it periodically. Reservations for a particular flow are initiated by its receiver; the messages that store state from the receiver to the sender are called RESV messages.

However, IP datagrams from the receiver to the sender do not typically follow the same path as datagrams in the opposite direction. Thus, an RSVP message (PATH message) from the sender to the receiver of the flow has to be sent prior to resource reservation in order to scout out which path the datagrams belonging to the flow will follow. PATH messages contain the path that RESV messages must traverse backwards towards the sender installing state in the router. Figure 3-21 shows an RSVP message flow.

The use of soft states also prevents the network from retaining unnecessary state information when routes change. If the route towards the flow destination changes as a consequence of the routing protocol operation, all of the datagrams will take a new route. Periodic PATH messages will also take the new route. Routers in the old route will not receive any more PATH messages, and therefore no RESV message will be received either. At that point, the state stored in those routers will time out and be deleted.

Differentiated Services (DiffServ)

We have seen that RSVP and the integrated services architecture provide different treatment for different flows in the routers. Routers filter packets based on the information received in RSVP messages. For instance, as long as flows are defined by the destination address and destination port number of the packets, routers have to examine the destination address and

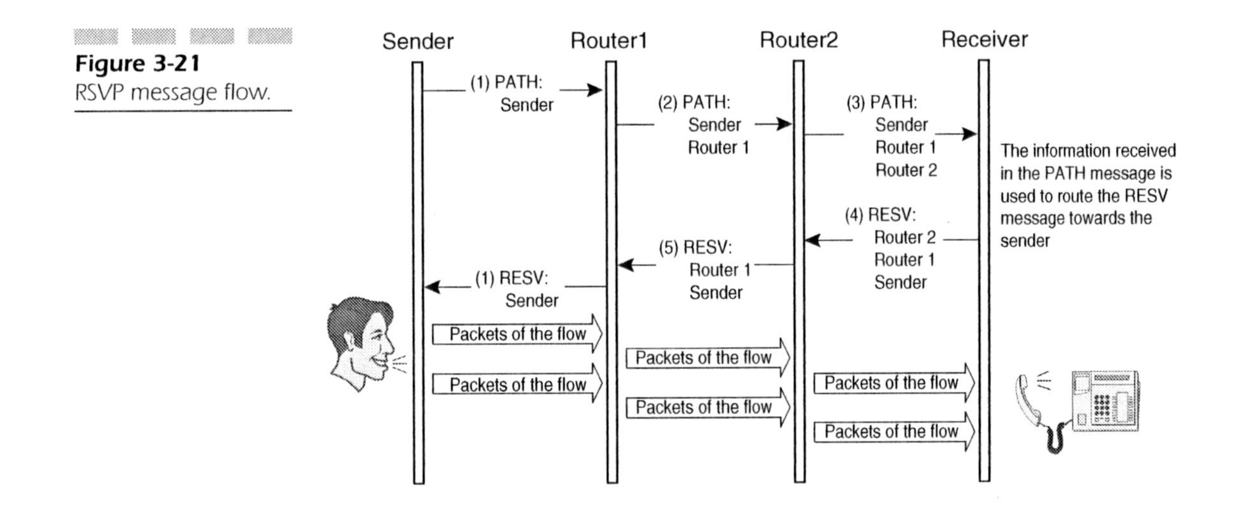

Figure 3-21
RSVP message flow.

port number of every datagram and map it to state information. The *differentiated services* (DiffServ) architecture [RFC 2475] simplifies the task by defining several traffic classes with different priority levels. Packets are tagged at the edge of the network with the required priority level. Routers in the network take their cue from these tags; each tag is associated with a particular way of handling packets (referred to as *Per-Hop Behavior* [PHB]), and the router only has to read the tag and look up its PHB. Examples of standard PHBs include *expedited* forwarding [RFC 2598] and *assured* forwarding [RFC 2597]. The former imitates the behavior of a circuit-switched network and the latter provides drop precedence.

DiffServ scales better than integrated services because it releases routers from the requirement to maintain per-flow state. Even with DiffServ, however, networks still need admission control mechanisms. Otherwise nothing would prevent end systems from tagging all traffic as high priority and swamping the network. RSVP can be used for admission control, so expect to see RSVP and DiffServ used together for maximum scalability, as shown in Figure 3-22.

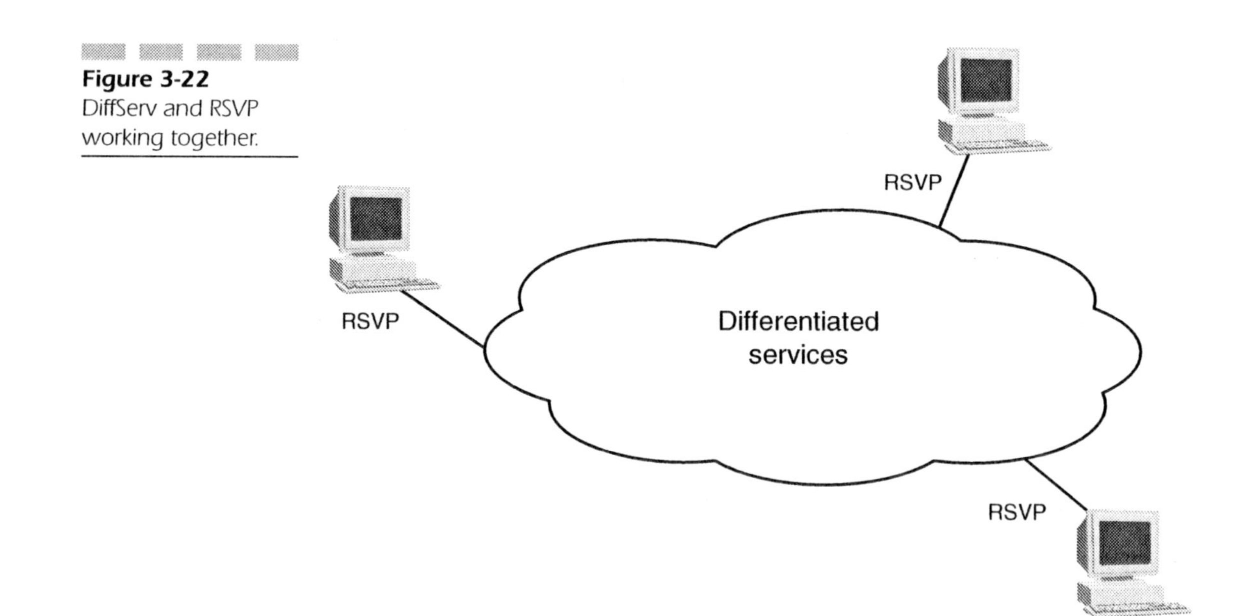

Figure 3-22 DiffServ and RSVP working together.

Session Announcement Protocol (SAP)

When I decide to watch TV, I usually check the program guide to see which channels are broadcasting something of interest. Once I've made a selection, I turn on the TV and turn to the correct channel. The program guide contains information about the contents of available programs, the channel they are on and the broadcast schedule.

The Internet utilizes a similar procedure. I need to select the most interesting multicast session from among all those available. I also need to know how to configure multimedia tools to receive the session chosen. It is important to know, for instance, whether a session consists of just audio or contains video as well.

The *Session Announcement Protocol* (SAP) [RFC 2974] comes into play to distribute information about multicast sessions among potential receivers. SAP undertakes the task of multicasting session descriptions on a well-known multicast address and port (Figure 3-23). Because multicast technology does not provide reliability, SAP announcements are unreliable and must be retransmitted periodically.

SAP announcements use a fixed amount of bandwidth, so a SAP host can afford to listen to all the announcements sent by other hosts on the same address and port. Depending on the number of announcements, the host

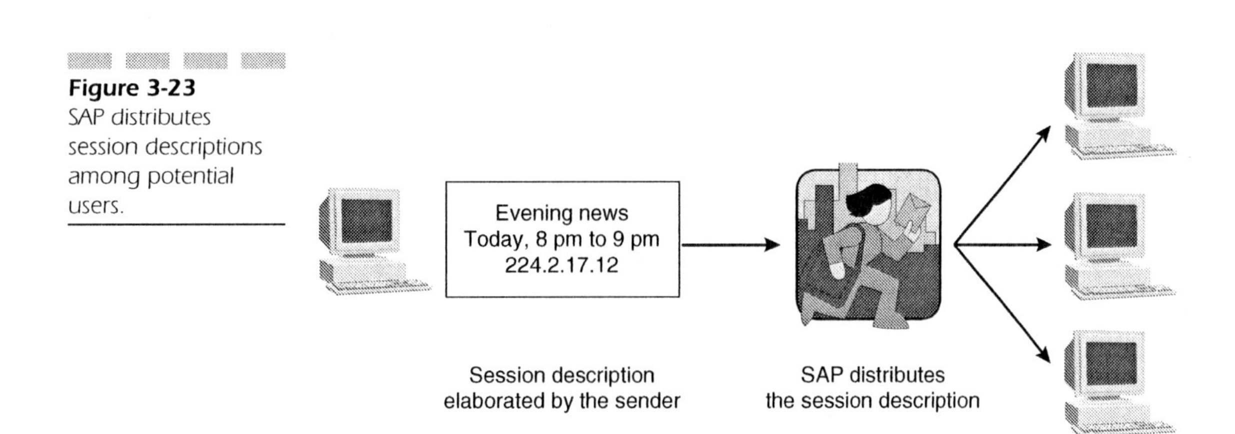

Figure 3-23 SAP distributes session descriptions among potential users.

chooses a retransmission rate for the announcement of its session. Hence, the more sessions that are present, the longer the interval becomes between retransmissions.

Finally, SAP announcements can be encrypted and can make use of authentication mechanisms. Encryption and authentication provide the requisite level of privacy and check the identity of the creator of a particular session.

Session Descriptions

Although SAP multicasts session descriptions to potential receivers, it does not define the format of those descriptions. Following the IETF way, various protocols besides SAP can be used to describe sessions. (For recommended format, consult the *Session Description Protocol* (SDP) [RFC 2327].) SAP announcements carry all description formats, but have no way to negotiate the protocol used for describing sessions. Therefore, when a system fails to understand an SAP announcement for some reason, it has no recourse. It cannot ask to receive the same announcement with a different Session Description Protocol inside. Instead, SDP serves as the common protocol for session description, and all applications must support SDP.

Session Description Protocol (SDP)

Session Description Protocol (SDP) [RFC 2327] specifies how the information necessary to describe a session should be encoded. SDP does not include any transport mechanism or any kind of parameter negotiation. An SDP description is simply a chunk of information that a system can use to join a multimedia session. It includes, for instance, IP addresses, port numbers, and times and dates when the session is active.

Turning again to the program guide analogy, a session description in TV context would look like the following: "Tune in channel 5 at nine o'clock this evening to see a soccer match," or "Turn on channel 2 at nine o'clock every evening to watch the news." Just as a session description in the TV context should contain information about how to receive the broadcast session (channel 5), when the session is broadcast (from nine o'clock to approximately eleven o'clock), and information about the contents of the session (soccer).

We can get information about our favorite TV programs in many ways. We can read about a program in the newspaper or in the TV guide. We can consult the teletext or we can even receive a telephone call from a friend telling us about a program. However, no matter how we find out about our favorite program, we always need the same information: which channel, when, and what program.

The information needed in the Internet context to receive a multimedia session is slightly different, but the concept is the same. SDP can be used to describe sessions no matter how session descriptions are distributed. They can be distributed using SAP, or they can be sent inside an e-mail. They can also be found on a Web page and retrieved using a normal Internet browser. We will shortly discover that SIP also carries session descriptions in its messages.

SDP Syntax

It is worthwhile to spend some time studying what SDP session descriptions look like because they appear in many SIP messages. SDP session descriptions are text-based, as opposed to a binary encoding such as ASN.1. An SDP session description consists of a set of lines of text of the form:

```
Type = value
```

The type field is always one character long and the format of the value field depends on which type it applies. An SDP description contains session-level information and media-level information. The session-level information applies to the whole session. It can be, for instance, the originator of the session or the session name. The media-level information applies to a particular media stream. It can be, for instance, the codec used for encoding the audio stream or the port number where the video stream is headed.

An SDP session description begins with the session-level information and the media-level information, if any is present, comes after. The session-level section always starts off with v=0, where *v* is the type and 0 is the value. This line indicates that the protocol version is zero (SDP version 0).

Ensuing lines, up until the first media-level section or the end of the session description as the case may be, provide information about the whole session.

Media-level sections begin with an *m* line. The lines below it, until the next *m* line occurs or until the end of the session description, provide information about that particular media stream. The following is an example of an SDP session description:

```
v=0
o=Bob 2890844526 2890842807 IN IP4 131.160.1.112
s=SIP seminar
i=A Seminar on the Session Initiation Protocol
u=http://www.cs.columbia.edu/sip
e=bob@university.edu
c=IN IP4 224.2.17.12/127
t=2873397496 2873404696
a=recvonly
m=audio 49170 RTP/AVP 0
a=rtpmap:0 PCMU/8000
m=video 51372 RTP/AVP 31
a=rtpmap:31 H261/90000
m=video 53000 RTP/AVP 32
a=rtpmap:32 MPV/90000
```

In this example, the session-level section consists of the first nine lines, from v=0 to a=recvonly, and has three media-level sections: one audio stream and two video streams. The *o* line indicates the creator of the session (in this case, Bob) and the IP address of his site. The *s* line contains the name of the session and the *i* line contains general information about the session. The *u* line provides a *Uniform Resource Locator* (URL) where more information about the topic of the session can be retrieved. The *e* line contains the e-mail address of the contact person for this session. The multicast address where the session can be received is described in the *c* line and the *t* line indicates when the session is active. The last line of the session-level section, the *a* line, indicates that this is not an interactive session; it's receive only. The format of the *m* lines is especially important. An *m* line begins with the media type. In the previous example, the media types are audio for the first media stream and video for the second and the third.

```
m=<media type> <port number> <transport protocol> <media formats>
```

The port number indicates where the media can be received. The transport protocol field usually takes the value of RTP/AVP, but can take another value if a protocol other than RTP is used. RTP/AVP refers to the audio/video profile for RTP; in this example, encoded audio and video are transported using RTP over UDP.

The media format depends on the type of media transported. For audio, it's the codec being used. In this example, a value of zero means that the audio is encoded in a single channel using PCM μ-law and sampled at 8 kHz.

The a=rtpmap lines convey information, such as the clock rate or number of channels, about the media formats used. In the second media stream of this example, the media format number 31 is referred to as H.261 and it uses a clock rate of 90 KHz.

Table 3-1 contains all the types defined by SDP and their meaning.

Extending SDP The media attribute lines, the *a* lines, provide a means to extend SDP. When an application needs a feature missing in SDP, it can add an *a* line containing it. For example, if the creator of a multicast session wanted the receivers to play the audio at a certain volume, he or she could define a new media attribute and add it at the end of the media-level section.

```
m=audio 49170 RTP/AVP 0
a=volume:8
```

Applications that understand this new *a* line will play the audio at volume 8. When an application finds an *a* line that it doesn't understand, it simply ignores the line and proceeds as if no line had been encountered. The application that failed to understand our new a=volume line could still receive media properly, although it would not be able to play back at the proper volume.

Table 3-1
SDP Types

v	Protocol version
b	Bandwidth information
o	Owner of the session and session identifier
z	Time zone adjustments
s	Name of the session
k	Encryption key
i	Information about the session
a	Attribute lines
u	URL containing a description of the session
t	Time when the session is active
e	E-mail address to obtain information about the session
r	Times when the session will be repeated
p	Phone number to obtain information about the session
m	Media line
c	Connection information
i	Information about a media line

For those of you particularly interested in this topic, the IETF is already evaluating some proposed extensions in the form of new *a* lines that will provide QoS when SIP and SDP are used together [draft-manyfolks].

SDP Next Generation (SDPng)

Originally, SDP was designed to describe multimedia sessions in the Mbone, but now it is finding use in many other contexts. Among others, SDP is used with the *Real-Time Streaming Protocol* (RTSP) [RFC 2326] for streaming services, with SIP for conference invitations, and for devices with a master/slave configuration using *Media Gateway Control Protocol* (MGCP) [RFC 2705] or H.248. Because SDP was not designed for working in all these environments, it's an imperfect fit and lacks features needed by some applications.

These new contexts, as well as a number of future applications that might need a session description mechanism, put new requirements [draft-kutscher-mmusic-sdpng-req] on the successor of SDP. Currently, this is called SDP next generation (SDPng) [draft-ietf-mmusic-sdpng] and is being developed in the *Multiparty Multimedia Session Control* (MMUSIC) working group. SDPng will try to provide richer session descriptions and a better means for capability negotiation than SDP. However, one of the key concepts behind the design of SDPng is simplicity. Therefore, the trade-off is that SDPng must provide a reasonable level of functionality within a reasonable level of complexity.

Real-Time Streaming Protocol (RTSP)

The Real-Time Streaming Protocol (RTSP) [RFC 2326] is used to control multimedia servers, typically for streaming applications. The use of RTSP between an end user and a multimedia server can be compared to the use of the remote control with a VCR. The user can tell the multimedia server, for instance, to initiate a certain audio/video stream using the play button, to freeze the stream at a particular moment using the pause button, or to begin replay at a certain position using the forward and rewind buttons. The user can also command the server to record certain media using the record button. RTSP can be used, for example, to implement a distributed answering machine, or to record an event that is being multicast on the Internet.

Usage Example of the Internet Multimedia Conferencing Toolkit

Let us see how some of the protocols analyzed in this chapter can be combined in order to multicast a film on the Internet. The end user controlling the multicast of the film elaborates a session description using SDP,

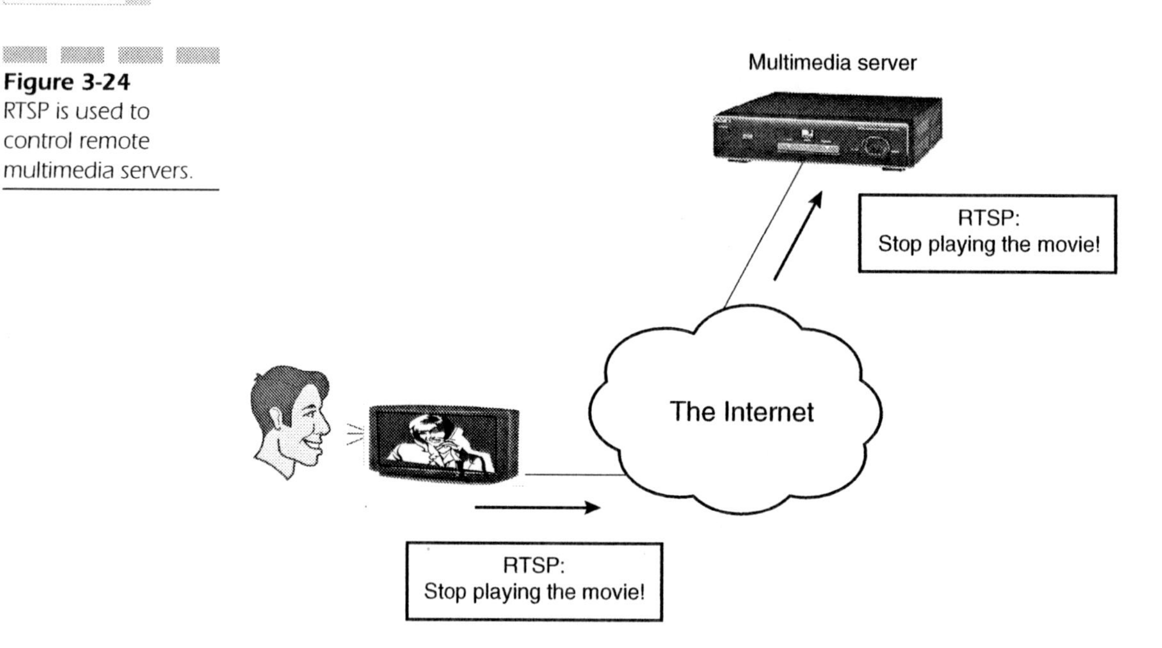

Figure 3-24
RTSP is used to control remote multimedia servers.

indicating when the film is going to be multicast, what the film is about, and the parameters needed to receive the media. These parameters will include multicast addresses, port numbers, and media formats at a minimum. This SDP session description is then distributed via SAP to potential receivers, making use of multicast routing.

Interested end users will examine the SDP they've received and configure their media tools properly to be able to watch the film at the appointed time. When the film is programmed, the session controller will use RSTP to alert the multimedia server where the film resides to begin multicast using the SDP session description that was previously distributed.

The media server will multicast RTP packets containing the audio and the video of the film. It will use RTCP to store statistics about the reception quality that the receivers are experiencing. RSVP might also be used to grant a certain QoS between media server and receivers.

The Session Initiation Protocol: SIP

This chapter focuses squarely on the *Session Initiation Protocol* (SIP). We describe the functionality SIP provides and the entities defined by the protocol. We analyze its good features and explain where SIP fits in the developing picture of communications. Finally, we explain why SIP is such a compelling example of good protocol design. This chapter explains what we can expect from SIP, but it does not go into protocol details. For information on protocol operation, including messages, syntax, and concrete use cases, turn to the next chapter. Separating protocol behavior from protocol operation enables the reader to distinguish between SIP's functionality and how this functionality is achieved.

SIP History

In the previous chapter, we saw that the Internet multimedia conferencing architecture encompasses many protocols. These were not all developed at the same time. Beginning with the first multimedia systems, this architecture was dynamically evolving. New protocols were designed and existing protocols were improved. As time went by, the Internet stretched to provide more and more multimedia services.

Yet this architecture still had a missing piece: it did not have a way to explicitly invite users to join a particular session. A multicast session could be announced using *Session Announcement Protocol* (SAP), for instance, but it was up to the potential receiver to check through all of the announced sessions periodically to find one he or she wanted to join. It was impossible for one user to inform another user about a session and invite him or her to participate.

Suppose I'm watching an interesting movie being multicast on the Mbone and I recall a friend who would probably also be interested in seeing it. I need a simple means to alert my friend, send her a description, and invite her to join the session (Figure 4-1).

Inviting users to Mbone sessions was the original purpose of SIP when the *Internet Engineering Task Force* (IETF) first commissioned it. The protocol has evolved steadily and SIP is currently used to invite users to all types of sessions, including multicast and point-to-point sessions.

SIP, as we know it today, was not designed from scratch, but resulted from the merger of two IETF protocols proposed for the same purpose. SIP picked up the best features from each protocol and from then on, all efforts within the community converged on it.

Figure 4-1
SIP enables us to invite users to sessions.

Session Invitation Protocol: SIPv1

Although the first voice transmissions over packet-switched networks took place around 1974, the first multimedia conference systems appeared in the early 1990s. Thierry Turletti developed the *INRIA Videoconferencing System* (IVS), a system for audio and video transmissions over the Internet. An IVS user could call another user and they could establish a unicast session. IVS could also be used in multicast sessions. Capitalizing on H.261 video coding over the Internet, the work on IVS was instrumental in developing the *Real-time Transport Protocol* (RTP) payload format for H.261 video streams [RFC 2032].

Soon thereafter, Eve Schooler developed the *Multimedia Conference Control* (MMCC). MMCC software provided point-to-point and multipoint teleconferences, with audio, video, and whiteboard tools.

To connect various users, MMCC used the *Connection Control Protocol* (CCP), a transaction-oriented protocol. A typical transaction consists of one request (from the user) and one response (from the remote user). For transport, CCP used *User Datagram Protocol* (UDP) so it implemented time-outs and retransmissions to ensure the reliable delivery of protocol messages.

These two first multimedia systems gave way to the design of the *Session Invitation Protocol* created by Mark Handley and Eve Schooler. The first version of SIP, SIPv1, was submitted to the IETF as an Internet draft on February 22, 1996. SIPv1 used *Session Description Protocol* (SDP) to describe sessions and UDP as a transport. It was text based.

The concept of registrations to conference address servers was prominent in SIPv1. Once a user had registered his or her location, an address server was able to route invitations to the proper user and also provide a certain level of user mobility. If someone were away from his or her normal workstation on business travel, for example, this user could choose to register his or her temporary workstation and receive invitations to local conferences.

Notably, SIPv1 only handled session establishment. Signalling stopped once the user joined the session and mid-conference controls were yet to come.

Simple Conference Invitation Protocol: SCIP

Also on February 22, 1996, Henning Schulzrinne submitted an Internet draft to the IETF specifying the *Simple Conference Invitation Protocol* (SCIP). SCIP was also a mechanism for inviting users to point-to-point and multicast sessions. It was based on *Hypertext Transfer Protocol* (HTTP) and thus utilized *Transmission Control Protocol* (TCP) as the transport protocol. Like SIPv1, it was text based. SCIP used e-mail addresses as identifiers for users, aiming to provide a universal identifier for both synchronous and asynchronous communications. SCIP signalling persisted after session establishment to enable parameter changes in ongoing sessions and closing existing sessions. Instead of recycling a mechanism for session description like SDP, it defined a new format for this purpose.

Session Initiation Protocol: SIPv2

At the 35th IETF meeting in Los Angeles, Schooler presented SIP and Schulzrinne presented SCIP. During this meeting and through the 36th IETF meeting, the usual colorful level of discussion ensued. Eventually, it was decided to merge the two protocols.

The resulting protocol kept SIP as a name, but changed the meaning of the acronym to *Session Initiation Protocol* and advanced the version to number 2 (Figure 4-2).

An Internet draft of SIPv2, authored by Mark Hanley, Schulzrinne, and Schooler, was submitted to the IETF in San Jose during the 37th meeting in December 1996. The new SIP was based on HTTP, but could use both UDP and TCP as transport protocols. It used SDP to describe multimedia

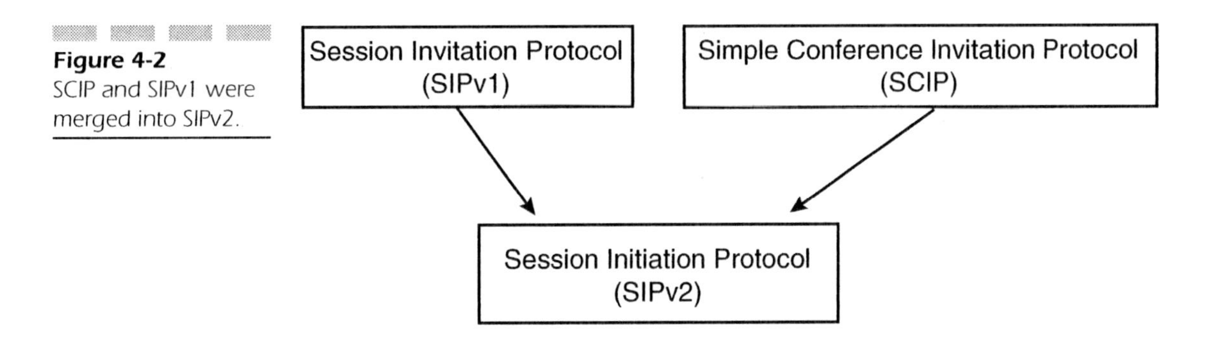

Figure 4-2
SCIP and SIPv1 were merged into SIPv2.

sessions, and it was text based. It remains the current version of the SIP protocol.

SIP development efforts were the province of the *Multiparty Multimedia Session Control* (MMUSIC) working group, chaired by Joerg Ott and Colin Perkins. The first draft grew out of the feedback received by the authors and discussions in the MMUSIC mailing list. In 1998, Jonathan Rosenberg was added as a co-author of the specification because he had contributed so much to the discussion, and in the following February (1999), SIP reached the proposed standard level and was published as RFC 2543.

As time went by, SIP gained importance in the IETF, resulting in the formation of a new SIP working group in September 1999. This working group was originally chaired by Joerg Ott, Jonathan Rosenberg, and Dean Willis. In August 2000, Brian Rosen replaced Rosenberg as co-chair because Rosenburg and Willis changed corporate affiliations to the same company, and diversity was desired among the chairs.

Close on the heels of the March 2001 IETF meeting, the SIP working group was split in two. Discussions about the main SIP specification and its fundamental extensions now take place in a group that continues to be called SIP, whereas discussions about applications that use SIP are carried out in a group called SIPPING. This division of labor should help manage the enormous number of contributions related to SIP that the IETF is being asked to consider.

Proposed Standard Status As of July 2001, SIP is still not a finished product. Its current status is proposed standard. Further review is needed before it can be awarded the next maturity level of draft standard, and it must have had at least two different interoperable implementations. Authors have received a large amount of review feedback from the community and the implementors' experiences are being recorded. All fixes and additions to the protocol are gathered in an Internet draft [draft-ietf-sip-rfc2543bis] destined to become the draft standard RFC when SIP is truly stable.

It is worth noting that although SIP is still a proposed standard, it is stable enough to be implemented in products. Developers hold periodic interoperability tests to ensure interoperability, originally known as *bake-offs* but now somewhat stodgily renamed *SIP interoperability events* at the behest of the baking industry (see http://www.cs.columbia.edu/ sip/sipit /pillsbury.html for the story). The first bake-off took place in April 1999 at Columbia University and now usually three SIP bake-offs take place in a year. SIP bake-offs have proved useful for finding bugs in the specification and elaborating solutions for them.

Functionality Provided by SIP

RFC 2543 describes the core of SIP: that is, the basic operation of the protocol. Besides this basic spec, a number of extensions to SIP have been defined in other RFCs and Internet drafts (Figure 4-3). Once again, this chapter will limit the discussion to the functionality provided by the basic specification.

Session Establishment, Modification, and Termination

SIP establishes, modifies, and terminates multimedia sessions. It can be used to invite new members to an existing session or to create brand new sessions. When SIP notifies my friend Bob that something he would find interesting is being multicast on the Internet, I'm invoking an existing session. However, if Bob calls up Laura to spread the news, this two-party call constitutes a new multimedia session with a single audio component. In this case, Bob is inviting Laura to join a session that has yet to be created. Additionally, it will only be created if two conditions are met: (1) Laura is willing to speak to Bob and (2) they can agree on the media parameters that will be used.

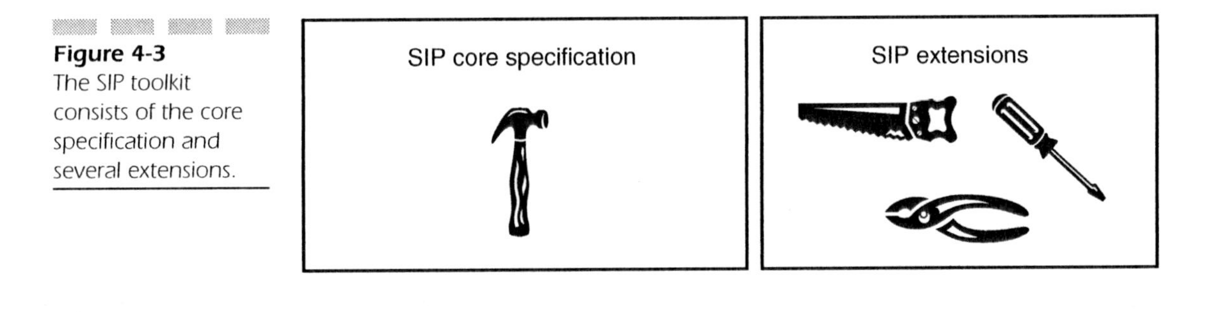

Figure 4-3 The SIP toolkit consists of the core specification and several extensions.

SIP is independent of the type of multimedia session handled and of the mechanism used to describe the session. It is equally useful for videoconferences, audio calls, shared whiteboards, and gaming sessions. Sessions consisting of RTP streams carrying audio and video are usually described using SDP, but some types of session can be described with other description protocols. Assuming Bob wants to play chess with Laura, he has the option of using a chess-specific session, which will be described by a chess-specific description protocol instead of SDP. If Bob and Laura were to play a video game over the Internet, they also would probably use a protocol other than SDP to describe their gaming session.

In short, SIP is used to distribute session descriptions among potential participants (Figure 4-4). Once the session description is distributed, SIP can be used to negotiate and modify the parameters of the session and terminate the session.

The following example illustrates all of these functions. Bob wants to have an audio-video session with Laura and plans to use a *Pulse Code Modulation* (PCM) codec to encode voice. In this example, the session distribution part consists of Bob sending Laura a session description with a PCM codec for the voice component of the session. Laura prefers to use a *Global System for Mobile Communications* (GSM) codec because it consumes less bandwidth, so she persuades Bob to do it her way. Both finally settle on a GSM audio codec, but the session cannot be established until this negotiation is concluded.

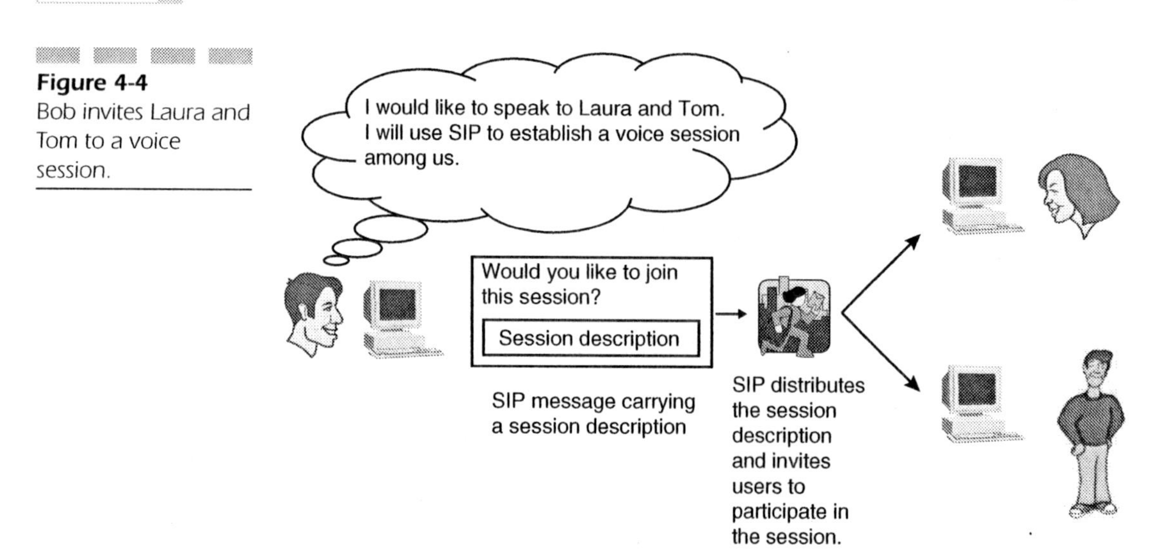

Figure 4-4
Bob invites Laura and Tom to a voice session.

Suddenly, in the middle of the audio-video session, Laura decides she is having a bad-hair day and wants to kill the video component. She modifies the session for audio only. When Bob then decides the conversation is over (I can't imagine why), the session is terminated.

Just as telephone systems inform a caller about the status of his or her call setup by playing different tones (busy tone or ringing tone), SIP provides the session initiator with information about the progress of his or her session setup (Figure 4-5).

User Mobility

SIP can't deliver a session description to a potential participant until he or she has been located. Frequently, a single user might be reached at several locations. For instance, a student using a computer room in the university typically works on a different workstation every day. Thus, he or she is reachable at different *Internet Protocol* (IP) addresses depending on which computer is available and wants to receive incoming session invitations only at his or her current location. Another person might want, for instance, to receive session invitations on his or her workstation in the morning when the user arrives at the office, on his or her desktop at home in the evening, and on his or her mobile terminal when the user is traveling.

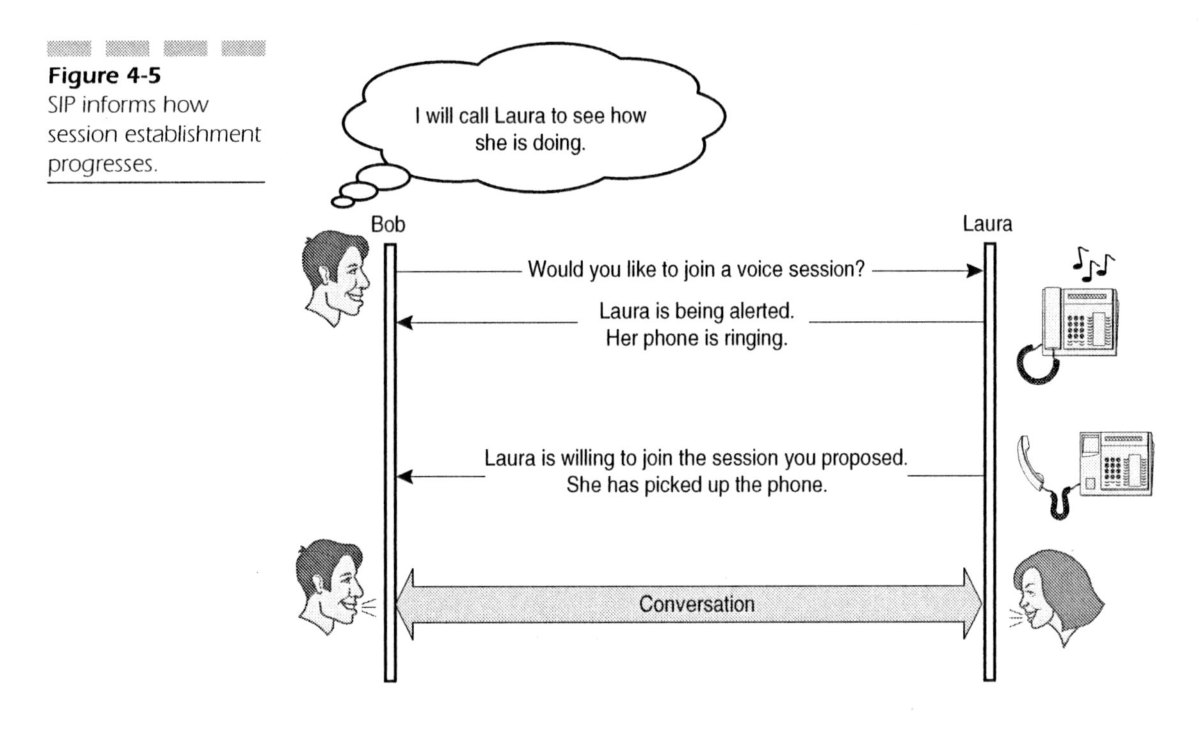

Figure 4-5 SIP informs how session establishment progresses.

SIP URLs We've already mentioned that SIP provides some user mobility. Users in a SIP environment are identified by SIP *Uniform Resource Locators* (URLs). The format of a SIP URL is similar to an e-mail address, generally consisting of a username and a domain name, which looks something like this: SIP:Bob.Johnson@company.com.

In the previous example, if we consult the SIP server that handles the domain company.com, we will find a user whose username is Bob.Johnson. Bob's URL might be SIP:Bob@131.160.1.112 instead, indicating that the host whose IP address is 131.160.1.112 has a user whose username is Bob.

Registrations We've noted that users register their current location to a server if they wish to be found. In this example, Bob is working on his laptop, whose IP address is 131.160.1.112. His login name is Bob. He registers his current position with the company server (Figure 4-6).

Now Laura wants to call Bob. She has his public SIP address (SIP:Bob.Johnson@company.com) because it's printed on his business card.

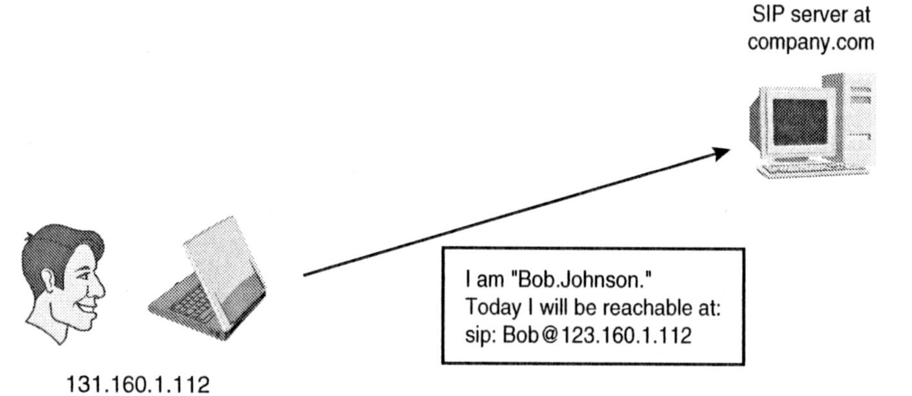

Figure 4-6
Bob registers his current position to the server.

So when the server at company.com is contacted and asked for Bob.Johnson, it knows where Bob.Johnson can be reached and a connection is made.

In this situation, SIP provides two modes of operation: redirect and proxy. In the proxy mode of operation, the server contacts Bob at 131.160.1.112 and delivers Laura's session description to him (Figure 4-7).

In the redirect mode, the server tells Laura to try SIP:Bob@131.160.1.112 instead (Figure 4-8).

A user may very well register several locations at one server. Or the user might register his or her locations with several servers. It's not unusual for various servers and locations to be contacted before a user is finally reached.

SIP Entities

The SIP protocol defines several entities, and it's vital to understand their role inside any architecture that uses SIP.

User Agents

User Agent (UA) is the SIP entity that interacts with the user. It usually has an interface towards the user. Say Bob wants to make a call over the Internet with his computer. He launches the proper program that contains a SIP

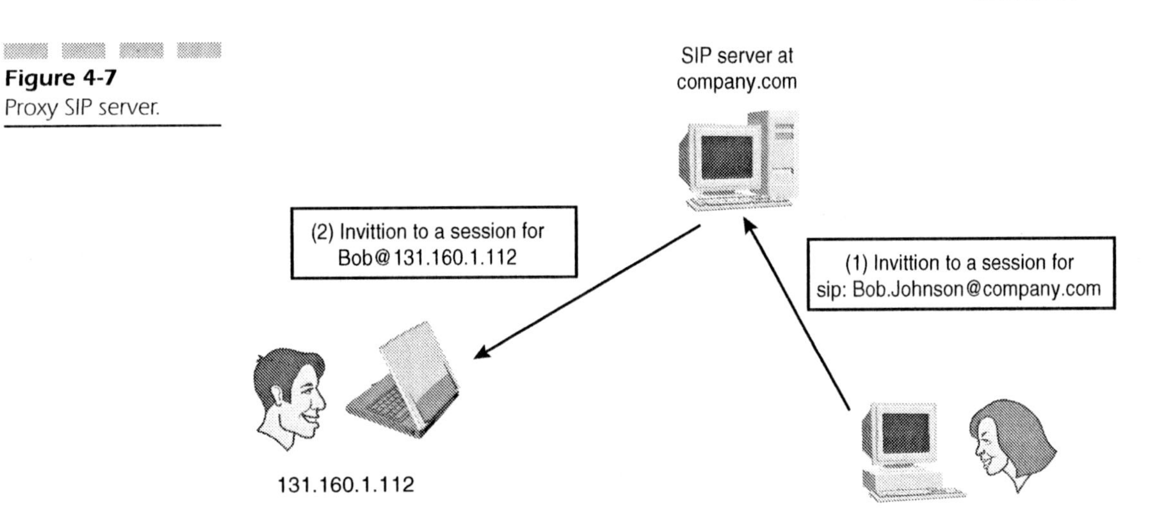

Figure 4-7
Proxy SIP server.

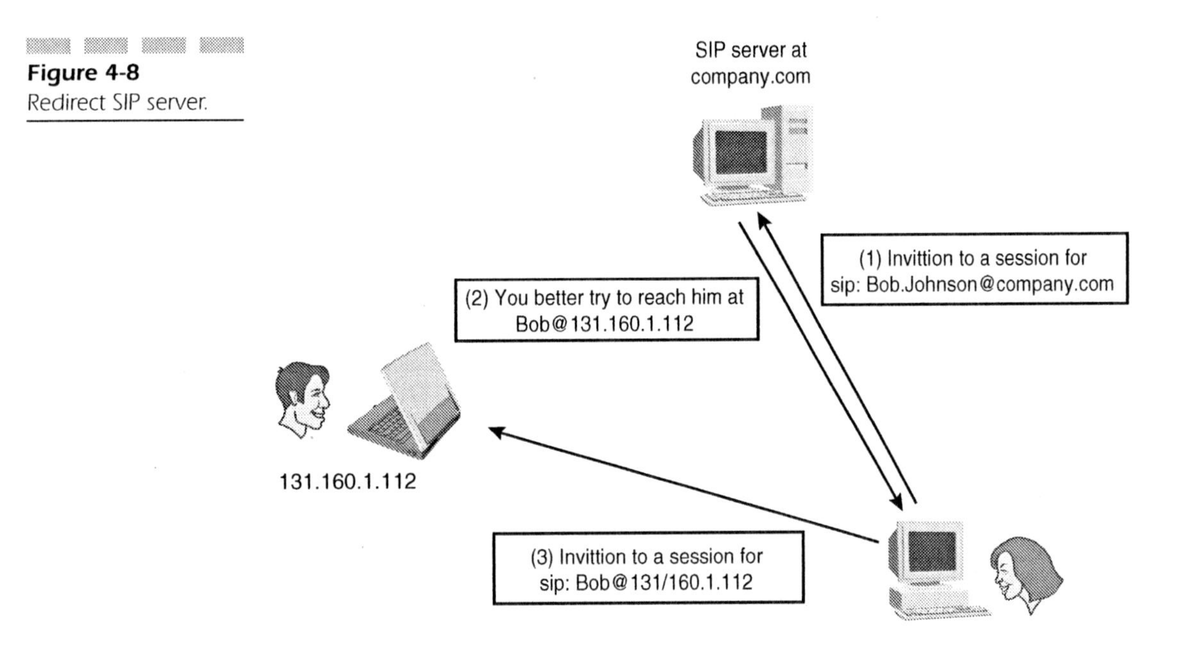

Figure 4-8
Redirect SIP server.

User Agent. The user interacts with the UA through the aforementioned interface—often a window with a selection of buttons. When Bob clicks the Call Laura button, the UA triggers the appropriate SIP messages to establish the call.

Laura also has a SIP UA in her computer. When its UA receives the invitation from Bob's, it alerts Laura by showing a pop-up window with two buttons: Accept call from Bob and Reject call from Bob. Depending on which button Laura clicks, her UA sends SIP messages back to Bob's UA. *All* interactions between users and the SIP protocol are mediated by UAs.

However, keep in mind that some systems using SIP are not directly connected to users. For example, Bob can redirect all session invitations received from midnight to 7 A.M. to his SIP answering machine. The machine will automatically establish sessions in order to record messages. It also contains a UA—one that does not necessarily maintain interaction with the user, but can still respond to invitations or forward invitations on Bob's behalf.

The lowly wake-up call is a good example of a session created automatically. The UA at the hotel reception is programmed to call the guest's UA at time *t*.

Media Tools All in all, SIP delivers a session description to a SIP UA. If the session described is a voice session, the UA will have to deliver it to the voice tool that will handle the audio. For other types of sessions, the UA will deliver the session to the proper media tool.

SIP UAs are sometimes incorporated into the same user interface with media tools for the session. An audio/video session can't be established without a SIP UA, an audio tool, and a video tool. If these three are combined under the same user interface, they appear as a single application to the user: a videoconference application.

The separation between the SIP UA handling the delivery of a session description and the media tools actually handling the contents of the session description is powerful. This separation enables SIP to establish any type of session.

What Does a SIP User Agent Look Like? SIP UAs are implemented on top of many different systems. They can run, for instance, in a computer as one among many applications, or they can be implemented in a dedicated device, such as a SIP phone. The device type will not affect SIP. Media tools might vary from device to device depending on the type of sessions invoked, but the SIP behavior is always the same.

Nonetheless, from the user point of view, SIP devices can look very different from each other. This is because the user interface varies with the kind of device. The user interface of a videoconference program running on a computer will most likely be a window with a selection of buttons to click, but a SIP phone will probably resemble a traditional telephone with the buttons 0 through 9, *, and #. SIP devices range from powerful computers accessing the Internet with a high-bandwidth connection to small devices with low-bit rate wireless connections. Figure 4-9 shows some examples.

I should mention that work to adapt SIP for household appliances is ongoing. Therefore, future examples of devices with SIP User Agents could include refrigerators, toasters, and lamps.

We are focusing on telephony examples because they are easier to understand and more immediately relevant for most readers; however, remember that SIP is powerful precisely because it can be used to establish any kind of session. Voice sessions are just one example.

Figure 4-9 Examples of devices that have SIP user agents.

Redirect Servers

Redirect servers help locate SIP UAs by providing alternative locations where the user can be reachable. For example, Laura wants to call Bob. On her monitor, Laura clicks the button that says Call Bob. Her UA first tries Bob's public address, but the domain company.com has a SIP redirect server handling incoming invitations. Instead, Laura's UA contacts this redirect server. The re-direct server knows that Bob can be located at SIP:Bob@131.160.1.112 when he is working at his office or at SIP:Bob@university.com when he is writing his dissertation. Thus, the redirect server will recommend that Laura's UA try SIP:Bob@131.160.1.112 and SIP:Bob@university.com rather than SIP:Bob.Johnson@company.com. The redirect server also has the capacity to prioritize and can tell Laura's UA that Bob is *most likely* to be reached at school rather than at work.

After being informed, Laura's UA tries Bob on both recommended SIP addresses. Note that a redirect server does not always return the address of the UA where the user actually is; it may just as easily return the address of another server with more knowledge about Bob's location instead (Figure 4-10).

This example shows that a redirect server does not initiate any actions to locate a user, but merely returns a list of possible locations where the user might be. The UA makes all of the attempts to locate the user. In this example, it is Laura's UA trying all possible locations until it finds Bob; this is the main difference between a redirect server and a proxy server. Proxy servers make subsequent attempts for the user rather than sending new contact information to the user.

Group Addresses Redirect servers can also be used to implement group addresses. To see how this works, assume that the public address for Company A's support department is SIP:support@company.com. Because this department has to give support around the clock, several people are always at work. Bob works from 8:00 A.M. until 4:00 P.M., Peter works from 4:00 P.M. until midnight, and Mary works from midnight until 8:00 A.M.. The redirect server at company.com is able to return different addresses depending on the time of the day so that if it receives a call for SIP:support@company.com at noon, it automatically returns SIP:Bob.Johnson@company.com.

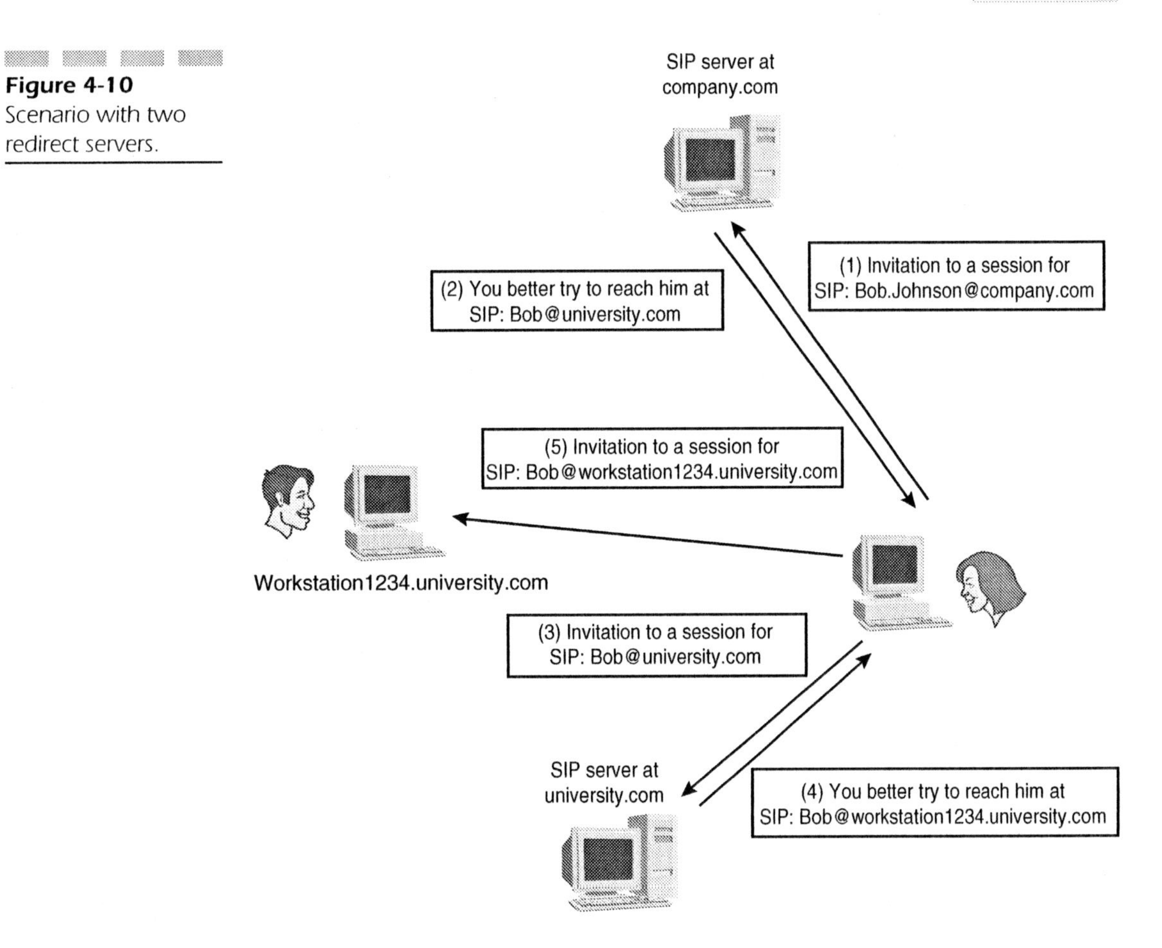

Figure 4-10
Scenario with two redirect servers.

Proxy Servers

Let us assume now that the domain company.com has a proxy server handling incoming invitations. When Laura's UA tries SIP:Bob.Johnson@ company.com, it will reach the proxy server at company.com, which will promptly try SIP:Bob@university.com on behalf of Laura's UA. If domain university.com also has a proxy server, it will try SIP:Bob@workstation1234.

university.com, where Bob is finally reached. In this scenario, Laura's UA tries only one location, but several proxies are in the path between UAs (Figure 4-11).

Forking Proxies When a proxy server tries more than one location for the user, it is said to *fork* the invitation. Forking proxies can perform parallel or sequential searches depending on their configuration. A parallel search consists of trying all of the possible locations at the same time, whereas a sequential search consists of trying each location individually.

Group Addresses Proxy servers also create group addresses. Figure 4-12 shows a forking proxy receiving an invitation for SIP:sales@company.com and trying all persons in the sales department until it finds one who is available.

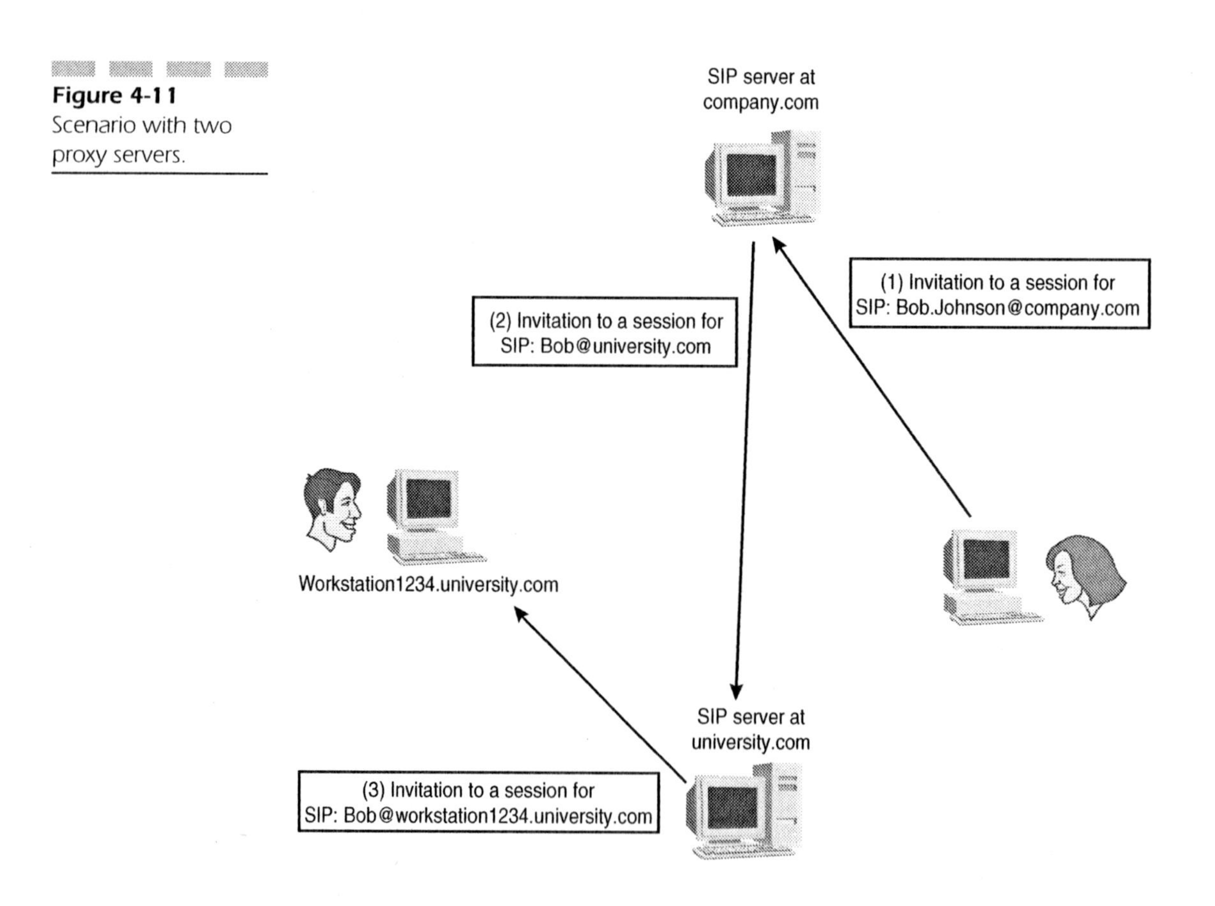

Figure 4-11 Scenario with two proxy servers.

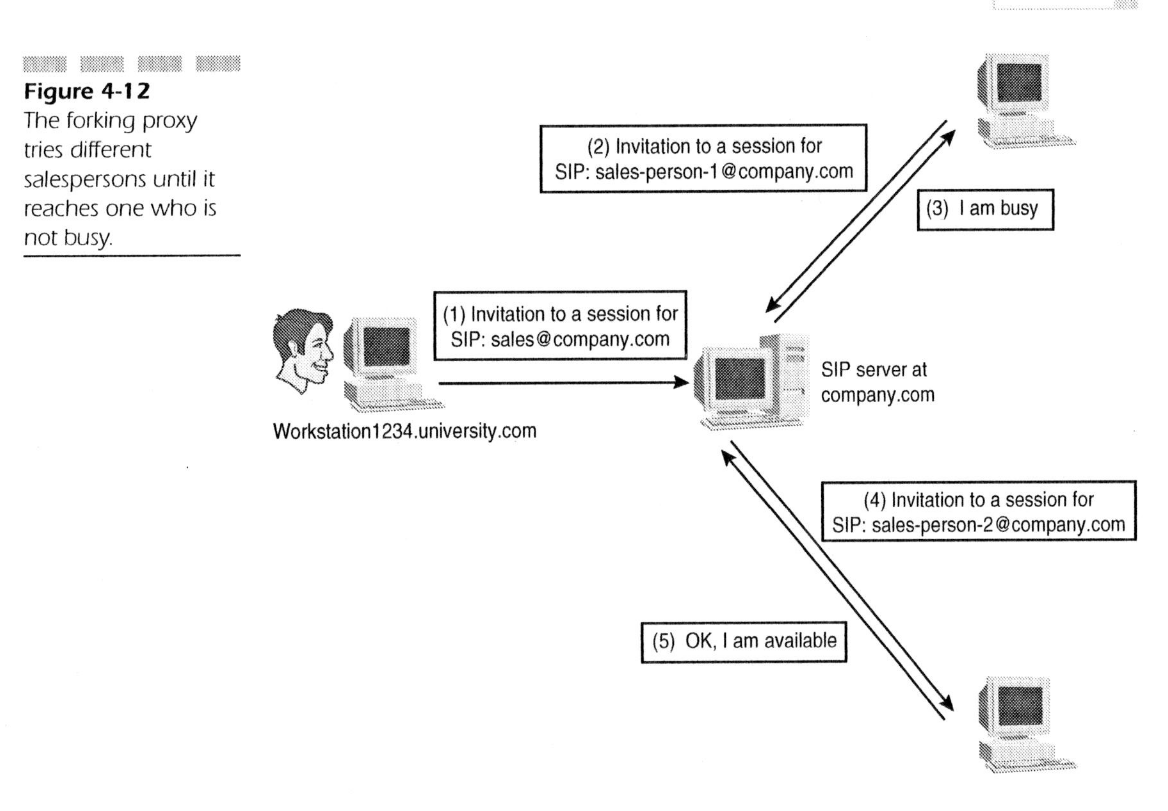

Figure 4-12 The forking proxy tries different salespersons until it reaches one who is not busy.

During session establishment, it is not uncommon for both kinds of servers (proxies and redirects) to be involved. The general term *SIP server* refers to both kinds of servers without differentiating on the basis of behavior. Actually, the same SIP server can act as a redirect or as a proxy depending on the situation. For instance, a SIP server can redirect all session invitations received for certain individuals and proxy the rest.

Registrars

Registrar refers to a SIP server accepting registrations. A registrar is usually co-located with a redirect server or a proxy server (Figure 4-13).

Location Servers

Location servers are not SIP entities, but they are an important part of any architecture that uses SIP. A location server stores and returns possible

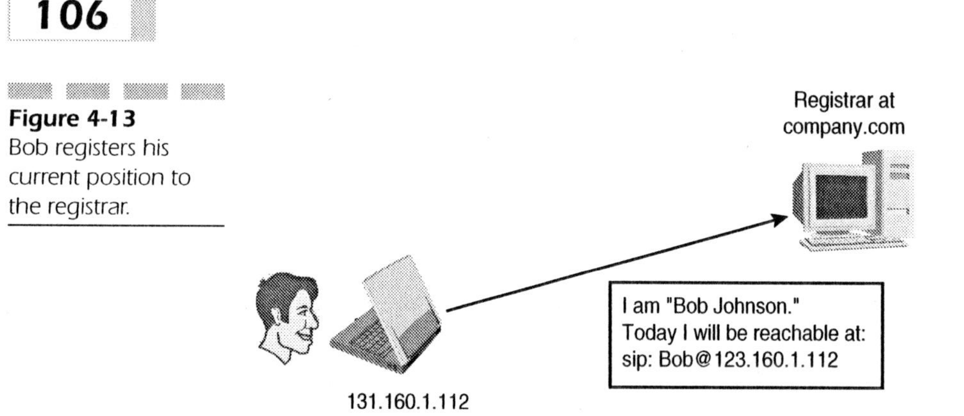

Figure 4-13 Bob registers his current position to the registrar.

locations for users. It can make use of information from registrars or from other databases. Most registrars upload location updates to a location server upon receipt. Figure 4-14 shows how this is done. In Figure 4-15, the proxy server at company.com consults a location server for a SIP URL where Bob might be reachable. The location server can provide the server with this information because the registrar previously uploaded it.

However, SIP is not used between location servers and SIP servers. Some location servers use *Lightweight Directory Access Protocol* (LDAP) [RFC 1777] to communicate with SIP servers.

Good Features of SIP

We have seen what functionality can be expected from the core SIP specification and which entities it defines. Now let's look at what makes SIP different from and better than many other protocols.

SIP Is Part of the IETF Toolkit

The IETF designed SIP with the Internet paradigm in mind. As a tool in the IETF toolkit, it performs its role and then takes advantage of other Internet mechanisms to perform additional tasks. This provides great flexibility because systems using SIP in conjunction with other Internet protocols can be upgraded in a modular way.

Figure 4-14 Registrar uploading information to a location server.

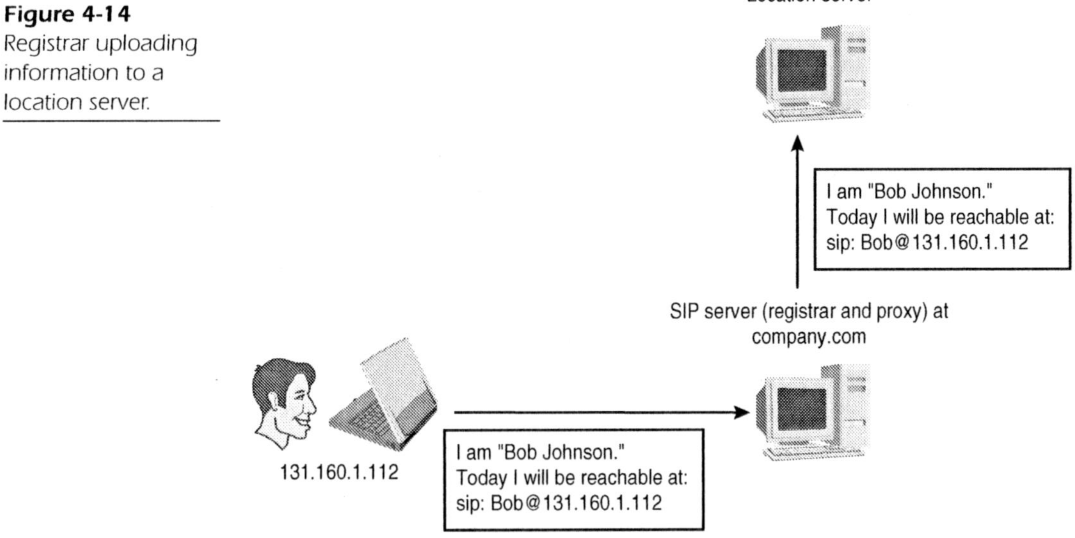

Figure 4-15 Proxy server consulting a location server.

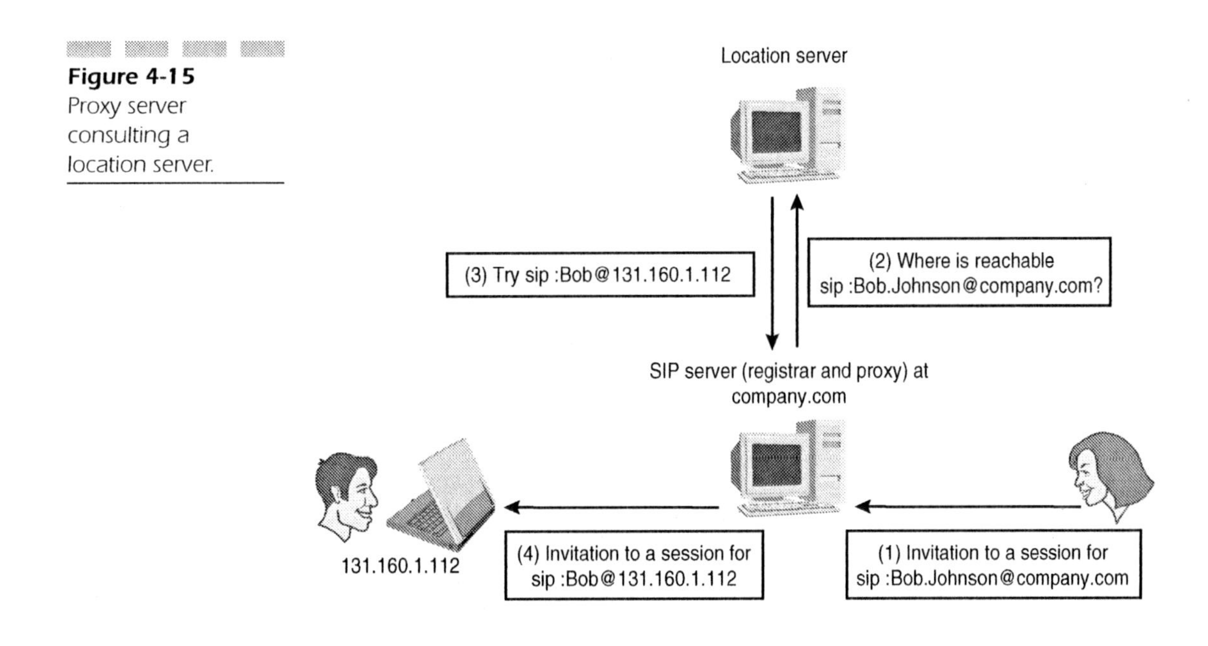

For instance, if a new authentication mechanism is proposed in the IETF, SIP systems can use it without implementing SIP modifications. A perfect example is the work on *SDP next generation* (SDPng) that the MMUSIC working group is undertaking. When SDPng is finalized, today's SIP systems will be able to carry SDPng session descriptions instead of the SDP descriptions that are carried today. SIP will be able to take advantage of new *Quality of Service* (QoS) mechanisms as well. We call that future-proof.

Separation Between Establishing and Describing a Session

SIP clearly distinguishes between session establishment and session description. As part of session establishment, SIP locates users as bidden, but it is silent on the topic of what users can do once the session is established. It does not define how a session should be described or session types. SIP just provides connectivity; what users do with it is outside of the scope of SIP.

This distinction makes SIP essentially cooperative. It can now be used together with SDP, for instance, to establish *Voice over IP* (VoIP) sessions, and it will soon be possible to combine SIP with new session description protocols to establish types of sessions that do not yet exist.

The concept we're describing is not dissimilar to the one that characterizes the IP layer. We saw that the most valuable service provided by the Internet is IP connectivity. All the rest are implemented taking IP connectivity as a base. Again, how IP connectivity is used falls outside the scope of IP itself. Hence, IP service creation is modular, fast, and efficient. SIP does not even assume that the session it has established will take place in the Internet. For instance, if Bob wants to invite Laura to join a conference call that is taking place in the *public-switched telephone network* (PSTN), all he has to do is use SIP to deliver the telephone number that she can dial to join in. In this example, the session description would contain a telephone number instead of IP addresses and UDP ports. When SIP delivers the session description to Laura, she reacts as she does to any kind of ping with the tool she's been given.

SIP provides just enough information for the invitee to accept the invitation; it is an exemplary IETF specification in that it performs its task. When we need to describe a session established by SIP, we should use another protocol designed for that purpose other than SIP.

Intelligence in the End System: End-to-End Protocol

The IETF community believes that end-to-end protocols are better for providing end-to-end services and that IP is an end-to-end protocol. IP provides connectivity between end points separated by a network of intervening routers. The routers perform the well-defined task of routing datagrams as efficiently as possible. Similarly, SIP provides connectivity between users with SIP servers. SIP servers have an equally well-defined task: routing SIP requests based on the Request-URI and responses based on Via headers.

SIP servers do not process session descriptions carried in SIP bodies because they don't need to in order to route SIP messages. This makes SIP an efficient protocol, along with the fact that all intelligence in a SIP network is located in the end systems—the UAs. SIP servers can be virtually stateless and forget everything about the transactions they're moving because the information needed to route a SIP message is contained in the message itself.

Interoperability

SIP is designed so that any implementation of the core protocol can interoperate with any other implementation and incorporates methods for negotiating the extensions that will be used in a session. Two highly advanced SIP UAs establishing a session are likely to use many extensions and sophisticated features. Yet if one of these advanced UAs needs to establish a session with a rudimentary UA, it always can. All SIP extensions are designed to be modular so that their use can be individually negotiated. I can pick one particular set of extensions for the first session I initiate and a completely different set for the next one.

Negotiation guarantees real interoperability between all the SIP users in the network. That's something new in many voice applications where protocols (*ISDN User Part* [ISUP], for instance) may have many incompatible flavors, necessitating the implementation of gateways between networks that fail to speak the same ISUP. Any implementation of gateways for protocol translation is undesirable because it breaks the end-to-end model and because some features present in a protocol flavor can be lost in the translation process. In contrast, SIP is a genuinely global protocol.

Scalability

SIP pushes the intelligence to the end system and obviates the need to store state information inside the network during a session. Once a user is located at session establishment, end-to-end communication is possible between end systems without the server's assistance. Servers that do not need to monitor signalling for the duration of the session can handle a larger number of sessions.

Some State in the Network In the next chapters, we'll see how SIP servers can be classified by the amount of state information that they store. But even when stateful, SIP servers that store state are used in the periphery of the network, leaving stateless SIP servers in the core of the network, where they have to handle a larger number of sessions. In part, SIP networks are highly scalable because they shift stateless operation to the points where the network is stressed.

SIP as a Platform for Service Creation

This is undeniably the most important feature of SIP. All of the SIP features explained so far are useful only insofar as they convert SIP into a good platform for user services.

Reuse of Components SIP (deliberately) makes use of many Internet components also exploited by other Internet applications. This makes SIP the perfect protocol to combine into different services for the user. Specifically, its similarities with HTTP [RFC 2068] and *Simple Mail Transfer Protocol* (SMTP) [RFC 821] make it easy to combine the most successful Internet services so far (Web and e-mail) with multimedia. SIP not only integrates services, but it also delivers them to the user's real location. For those in search of the holy grail of unified communications, SIP represents a revolution in combined services because SIP applications integrate Web browsing, e-mail, voice calls, videoconferencing, presence information, and instant messages in a straightforward way. Some people see SIP as the telecommunications industry's next killer app.

SIP Is Based on HTTP A SIP implementation is rather similar to an HTTP implementation for the obvious reason that the former is based on the latter. Both use a request/response model, both are text based, and both have a similar format for encoding protocol messages. These similarities enable an implementation to reuse code between the two protocols.

The utility in sharing code is highest for devices that have to provide Web browsing and SIP-based services. A SIP mobile phone with wireless Internet features will almost certainly implement both (Figure 4-16). Such devices lack the huge hard disks and tons of memory you're used to on the desktop; they're usually thin devices with serious footprint constraints for which the capacity gained by reuse is substantive.

SIP Uses URLs to Address SIP Resources Fortunately, the format used by SIP to address SIP entities is identical to the one used on the Web and by e-mail systems. This gives tremendous flexibility to SIP redirections and lets us integrate several forms of communication.

A redirect server typically returns an alternative address where the user can possibly be located. The redirect server is actually returning a URL. In most of our examples, it's a SIP URL, but the server has no problem returning a Web URL or an e-mail address.

Figure 4-16 Bob's terminal implements both SIP and HTTP.

This way Bob can configure his SIP redirect server to send incoming voice sessions to his e-mail account. The server obliges by returning mailto:Bob.Johnson@company.com in response to a SIP request (Figure 4-17). Laura, who's notoriously impatient, can choose between leaving Bob a voice mail or writing him an e-mail. If Bob has redirected incoming sessions on his Web page, she can amuse herself by looking at pictures of his new puppy, car, or house while her call is in a queue.

Web pages can also include SIP URLs besides e-mail addresses, providing click-to-dial features.

Having a SIP URL that can be redirected to any other means of communication considerably reduces the amount of different contact information needed by an individual. Currently, an average business card contains at least a fixed phone number, a mobile number, a fax number, and an e-mail address (Figure 4-18). SIP distills all this to a single URL.

Users will contact our SIP server specifying what type of service (voice call, e-mail, fax) they want, and our redirect server will provide them with the proper URL.

Same Routing Concept as SMTP SIP messages are routed in much the same way as e-mail messages. They can also carry multipart message bodies using *Multipurpose Internet Mail Extensions* (MIME) [RFC 2045]. However, SIP is not good at transporting large amounts of data; it's just not designed as a transport protocol.

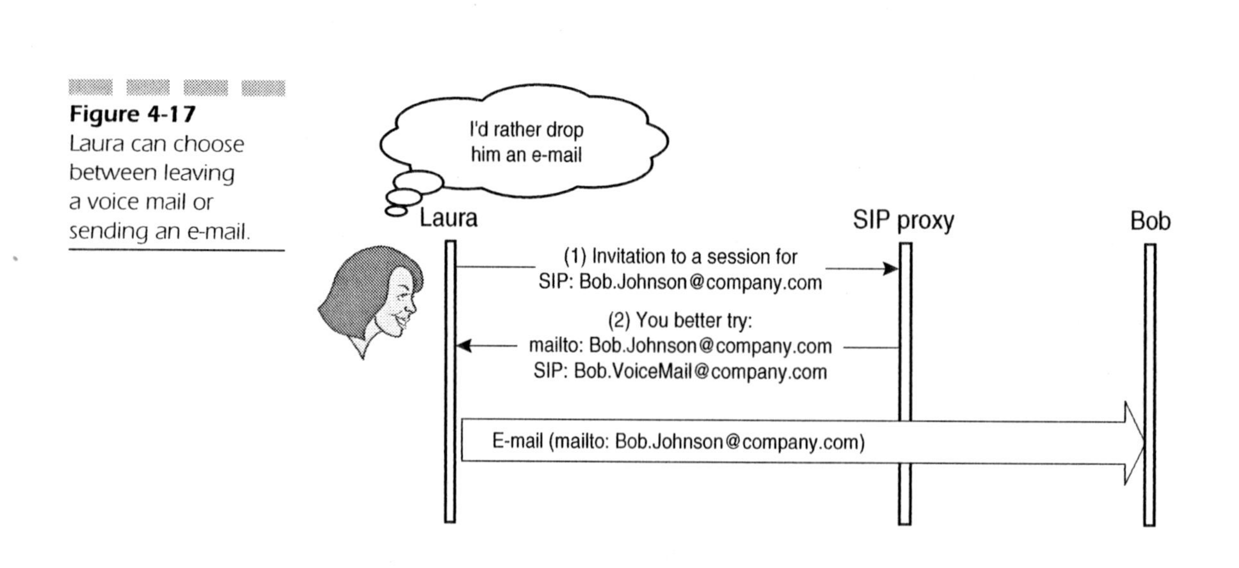

Figure 4-17 Laura can choose between leaving a voice mail or sending an e-mail.

Bob Johnson
Telephone: +1 212 555 5555
Mobile: +1 212 555 5556
Fax: +1 212 555 5557
E-mail: Bob.Johnson@company.com

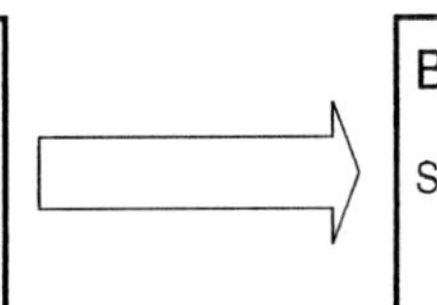

Bob Johnson
SIP: Bob.Johnson@company.com

Figure 4-18 Bob has a SIP business card.

On the other hand, it does deliver instant messages exceptionally well, which are small by definition and probably urgent, and intended to reach users at their present location. The same features also make SIP suitable as a protocol for presence. SIP registrars have to know when a user is online in order to deliver messages to him or her. Combining presence information with instant messages is another example of the advanced services that can be implemented by using SIP.

SIP Uses Existing Infrastructure for Providing New Services For instance, a service provider builds a SIP infrastructure for VoIP services, consisting of SIP proxies, redirect servers, and location servers. Customers of this service provider use SIP UAs to establish voice calls over the new infrastructure. Because VoIP is the service that the customers want, the service provider is successful.

But as time goes by, customers do not find VoIP that exciting any more, and what they really want to do is play interactive games. The service provider doesn't have to modify or retrofit at this point; it just has to procure an updated SIP UA that uses a gaming session description protocol to establish subscriber sessions instead of SDP. Because the SIP servers in the network ignore the contents of the session description, the entire SIP infrastructure built with VoIP in mind is instantly repurposed without any changes to the network.

Widespread Knowledge of How to Program SIP Applications Programming SIP applications requires certain skills that are already widespread among programmers. Most programmers are used to Web applications, text parsers, and scripting languages, which are exactly what come into play when coding SIP applications.

Besides, because SIP messages are human readable, no need exists for special protocol analyzers to discover why two different implementations do not interact properly. In fact, no special knowledge is needed in order to create a SIP service. Programmers fresh from the university are equipped with everything they need to innovate services, as long as they have the imagination. Not so long ago, this was far from true. Highly specialized knowledge was needed, and once the service was implemented, it was not easy to test. (Not everybody has the requisite telephone switch handy.) Today any personal computer can become a SIP server to test new applications.

This has an important implication. It means that the people creating a particular SIP service are the people who have the expertise on that particular service; that is, organizations creating gaming applications will come to understand gaming and organizations developing messaging applications will come to understand messaging. Today's programming practice (much bemoaned) is to find a programmer who knows a protocol and teach him or her how the applications should work. SIP enables people who know the functional needs of a certain community to create their own services.

SIP Enables Application Decomposition SIP enables us to combine simple applications into more complex services [draft-rosenberg-sip-app-components]. For instance, if I want to build an application that gets input in the form of a text in Spanish, translates it into English, and produces output in the form of speech, I can look at the solution as several simple applications working together. The first application performs text translation from Spanish to English. The second, which receives English text as input, converts it into English speech as output. This example comprises two application servers doing relatively simple things that can be tailored a hundred ways to meet a user's need. The user employs SIP signalling to coordinate both application servers in order to obtain the expected global result.

CHAPTER 5

SIP: Protocol Operation

This chapter describes SIP in further detail. It outlines how SIP functionality is achieved: that is, which messages are exchanged between different SIP entities and what their message formats are. We'll also examine multiple examples of how SIP works. These examples will usually consist of a message flow that gives a global picture of SIP operation.

However, I have also included one example where the reader can see all the messages in detail, including message headers, parameters, and session descriptions. Although it is not the purpose of this book to analyze all the protocol details, it is instructive to see, at least once, what SIP messages look like.

Client/Server Transactions

SIP is based on the Web protocol *Hypertext Transfer Protocol* (HTTP) and like HTTP, SIP is a request/response protocol. To understand the request/response mechanism used in SIP, we'll have to examine the following definitions of client and server.

A *client* is a SIP entity that generates requests. A *server* is a SIP entity that receives requests and returns responses. This terminology is inherited from HTTP, wherein a Web browser contains an HTTP client. When I type an address in my Web browser, such as http://www.accessmhtelecom.com, I am sending a request to a particular Web server. The Web server sends back a response with the information requested—namely, the Web page of McGraw-Hill's telecom publishing group.

SIP conforms to the same procedures. Following the same terminology, when two user agents exchange SIP messages, the *User Agent* (UA) sending requests is the *User Agent Client* (UAC) and the UA returning responses is the *User Agent Server* (UAS). A SIP request, together with the responses it triggers, is referred to as a SIP *transaction*.

SIP Responses

Upon reception of a request, a server issues one or several responses. Every response has a code that indicates the status of the transaction. Status codes are integers ranging from 100 to 699 and are grouped into classes, as shown in Table 5-1.

Table 5-1

SIP Response Classes.

Range	Response Class
100–199	Informational
200–299	Success
300–399	Redirection
400–499	Client error
500–599	Server error
600–699	Global failure

A response with a status code from 100 to 199 is considered provisional. Responses from 200 to 699 are final responses. A SIP transaction between a client and a server comprises a request from the client, one or more provisional responses, and one final response. (This rule has one exception, as we'll see in the next section.)

Together with the status code, SIP responses carry a reason phrase. The latter contains human-readable information about the status code. For instance, a status code of 180 means that the user invited to a session is being alerted. Therefore, the reason phrase might contain "Ringing." The reason phrase can, of course, be written in a language other than English because it will be read by a human. Accordingly, a computer processing a SIP response ignores the reason phrase. It finds sufficient information in the response code. However, it can display the reason phrase to the user, who will certainly find it more useful to know that the remote SIP phone is "Ringing" than to know a 180 response has been received. (From now on, we will cite responses using the status code followed by the reason phrase: for example, "180 Ringing.") Table 5-2 contains all the status codes currently defined with their associated default reason phrases.

SIP Requests

The core SIP specification defines six types of SIP requests, each of them with a different purpose. Every SIP request contains a field, called a *method*, which denotes its purpose. The list shows the six methods.

Table 5-2

SIP Response Codes

100	Trying	413	Request entity too large
180	Ringing	414	Request-URI too large
181	Call is being forwarded	415	Unsupported media type
182	Queued	420	Bad extension
183	Session progress	480	Temporarily not available
200	OK	481	Call leg/transaction does not exist
202	Accepted	482	Loop detected
300	Multiple choices	483	Too many hops
301	Moved permanently	484	Address incomplete
302	Moved temporarily	485	Ambiguous
305	Use proxy	486	Busy here
380	Alternative service	487	Request cancelled
400	Bad request	488	Not acceptable here
401	Unauthorized	500	Internal servere error
402	Payment required	501	Not implemented
403	Forbidden	502	Bad gateway
404	Not found	503	Service unavailable
405	Method not allowed	504	Gateway time-out
406	Not acceptable	505	SIP version not supported
407	Proxy authentication required	600	Busy everywhere
408	Request time-out	603	Decline
409	Conflict	604	Does not exist anywhere
410	Gone	606	Not acceptable
411	Length required		

- INVITE
- ACK
- OPTIONS
- BYE
- CANCEL
- REGISTER

We will see in subsequent chapters how some extensions to the core SIP specification define additional methods.

Both requests and responses can contain SIP bodies. The body of a message is its payload. SIP bodies usually consist of a session description.

INVITE INVITE requests invite users to participate in a session. The body of INVITE requests contains the description of the session. For instance, when Bob calls Laura, his UA sends an INVITE with a session description to Laura's UA. Let us assume that Bob's UA uses *Session Description Protocol* (SDP) to describe the session. Her UA receives the INVITE with the following session description:

```
v=0
o=Bob 2890844526 2890842807 IN IP4 131.160.1.112
s=I want to know how you are doing
c=IN IP4 131.160.1.112
t=0 0
m=audio 49170 RTP/AVP 0
```

The INVITE received by Laura's UA means that Bob is inviting Laura to join an audio session. From the session description carried in the INVITE, Laura's UA knows that Bob wants to receive *Real-time Transport Protocol* (RTP) packets containing Laura's voice on 131.160.1.112 *User Datagram Protocol* (UDP) port number 49170. Her UA also knows that Bob can receive *Pulse Code Modulation* (PCM) encoded voice. (RTP/AVP0 in the *m* line indicates PCM.)

Laura's UA begins alerting Laura and returns a "180 Ringing" response to Bob's UA. When Laura finally accepts the call, her UA will return a "200 OK" response with a session description in it.

```
v=0
o=Laura 2891234526 2812342807 IN IP4 138.85.27.10
s=I want to know how you are doing
c=IN IP4 138.85.27.10
t=0 0
m=audio 20000 RTP/AVP 0
```

At this point, Laura accepts the call and informs Bob that she will receive RTP packets on 138.85.27.10 UDP port 20000 (Figure 5-1).

If, when Laura and Bob are in the midst of the session, one of them wishes to modify the session, they just have to issue a new INVITE. This type of INVITE, called a re-INVITE, carries an updated session description. It might consist of new parameters such as port numbers for the existing media, or it might add new media streams. For instance, Bob and Laura can add a video stream to their voice conversation via a re-INVITE.

Significantly, SIP only handles the invitation to the user and the user's acceptance of the invitation. All of the session particulars are handled by the session description protocol used (SDP in this case). Thus, with a different session description, SIP can invite users to any type of session.

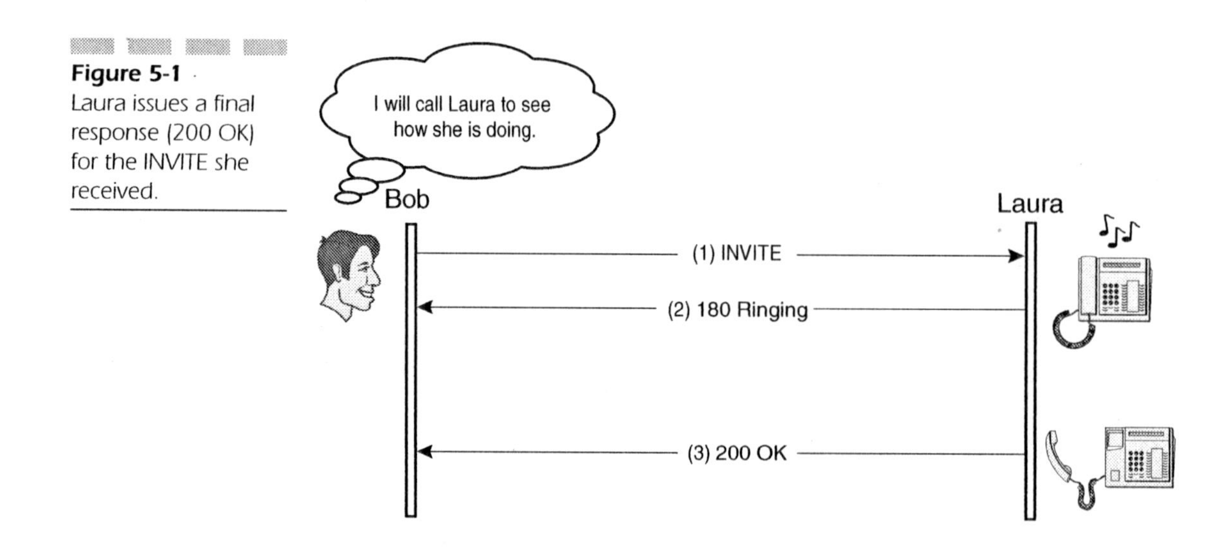

Figure 5-1 Laura issues a final response (200 OK) for the INVITE she received.

ACK ACK requests are used to acknowledge the reception of a final response to an INVITE. Thus, a client originating an INVITE request issues an ACK request when it receives a final response for the INVITE, providing a three-way handshake: INVITE-final response-ACK (Figure 5-2).

Why Does SIP Use a Three-Way Handshake? INVITE is the only method that uses a three-way handshake as opposed to a two-way handshake (METHOD-final response). Certain characteristics set the INVITE method apart from other methods. When a client issues a request other than INVITE, it expects a fast response from the server. However, the response from an INVITE request might take a long time. When Bob calls Laura, she may have to fish her SIP phone out of her coat pocket and press buttons, so the "200 OK" response that will come will be more or less delayed. Sending an ACK from the client to the server when the response is received lets the server know that the client is still there and that the session has been successfully established.

The three-way handshake also enables the implementation of forking proxies. When one of these forks a request, the client who issued the request will obtain several responses from different servers. Sending an ACK to every destination that has responded is essential to ensuring SIP operation over unreliable protocols such as UDP.

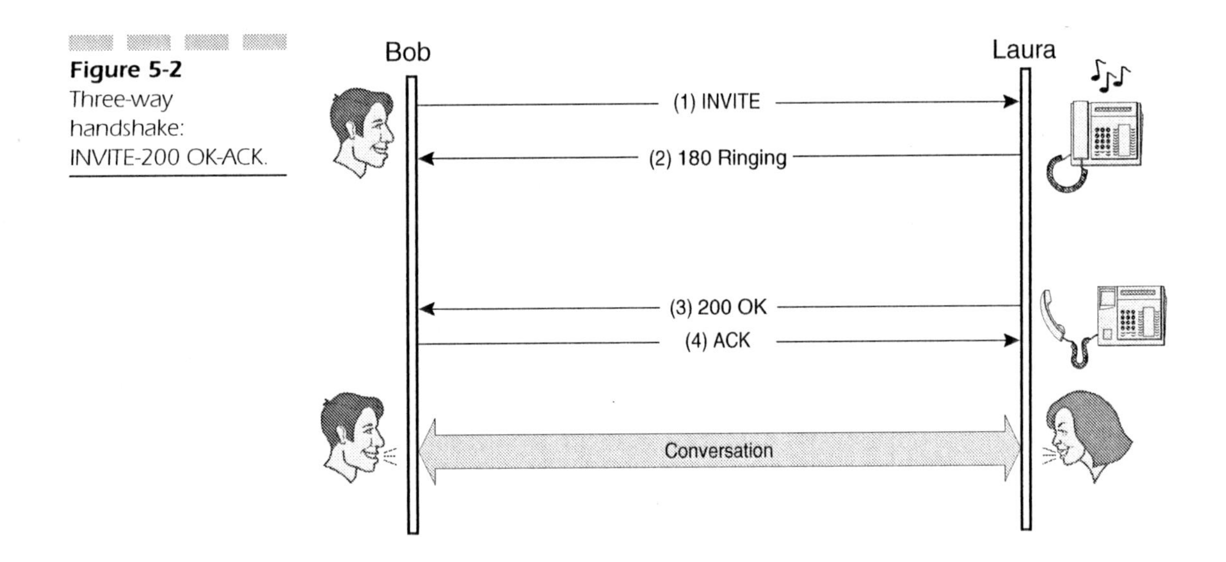

Figure 5-2 Three-way handshake: INVITE-200 OK-ACK.

Besides the speedy session setup and forking, INVITE's three-way handshake also enables us to send an INVITE without a session description, which will be sent later in the ACK. This feature is useful, for example, when SIP interworks with other signalling protocols that use different message sequencing.

However, the historical motivation for having a three-way handshake can be found in the old SIPv1 draft, in the section about how to provide reliable delivery of session invitation. The draft introduced the ACK method in order to avoid unsynchronized parties on session establishment, which might occur when a two-way handshake is used over an unreliable transport protocol such as UDP. Consider the following case where a two-way handshake is implemented.

Bob would send an INVITE to Laura and retransmit it until it received a final response from Laura. Until this final response is received, Bob cannot know whether Laura received the INVITE or it got lost in the network.

Bob waits for a while and because he gets no answer, he gives up and stops retransmitting the INVITE. Bob believes that no session has been established.

At roughly the same time, Laura accepts Bob's call and sends back a "200 OK" response. If this response gets lost, Bob will never receive it, so, Bob still believes that no session has been established. Because Laura observes that Bob has stopped retransmitting the INVITE, she assumes that Bob has received her 200 OK. Therefore, Laura thinks that the session has been successfully established (Figure 5-3).

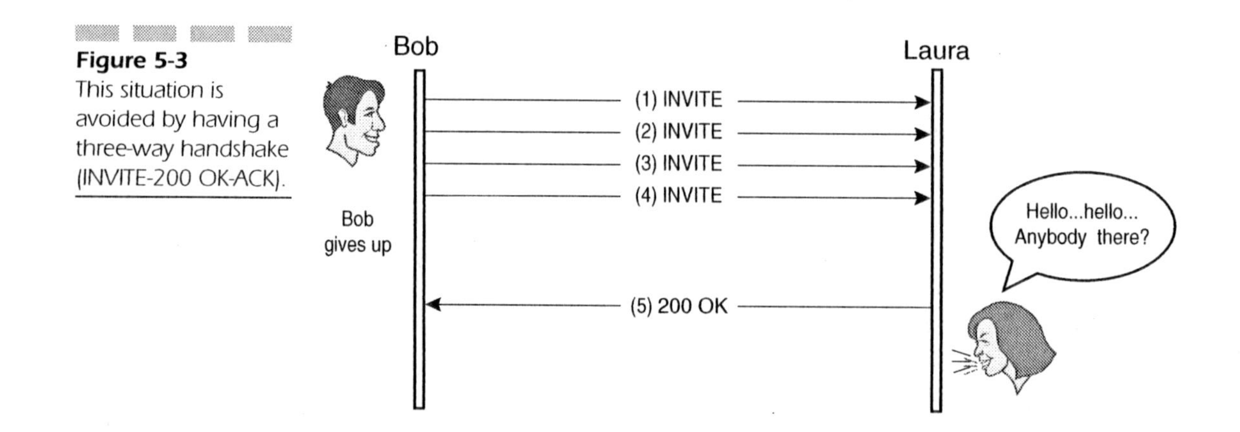

Figure 5-3 This situation is avoided by having a three-way handshake (INVITE-200 OK-ACK).

If a three-way handshake were in place for this scenario, Laura would not receive an ACK for her 200 OK response, given that Bob gave up some time ago. Thus, she would (correctly) think that the session was not established.

CANCEL CANCEL requests cancel pending transactions. If a SIP server has received an INVITE but not yet returned a final response, it will stop processing the INVITE upon receipt of a CANCEL. If, however, it has already returned a final response for the INVITE, the CANCEL request will have no effect on the transaction.

In Figure 5-4, Bob calls Laura and her SIP phone begins ringing, but nobody picks up for a while. Bob decides to hang up. He sends a CANCEL request for his previous INVITE. Upon reception of the CANCEL, Laura's SIP phone stops ringing. The server sends back a 200 OK response for the CANCEL, indicating that it was processed successfully.

It is important to remark that after the server has responded to the CANCEL request, it responds to the previous INVITE as well. It sends a "487 Transaction Cancelled" and the client finishes the INVITE three-way handshake by sending an ACK (INVITE-487 Transaction Cancelled-ACK). Therefore, the INVITE three-way handshake is always performed, even when the transaction is cancelled.

CANCEL requests are useful when forking proxies (proxies that issue more than one INVITE upon reception of just one INVITE) are in the path. When a forking proxy is performing a parallel search, it tries several locations at once. For example, a forking proxy knows of three possible locations

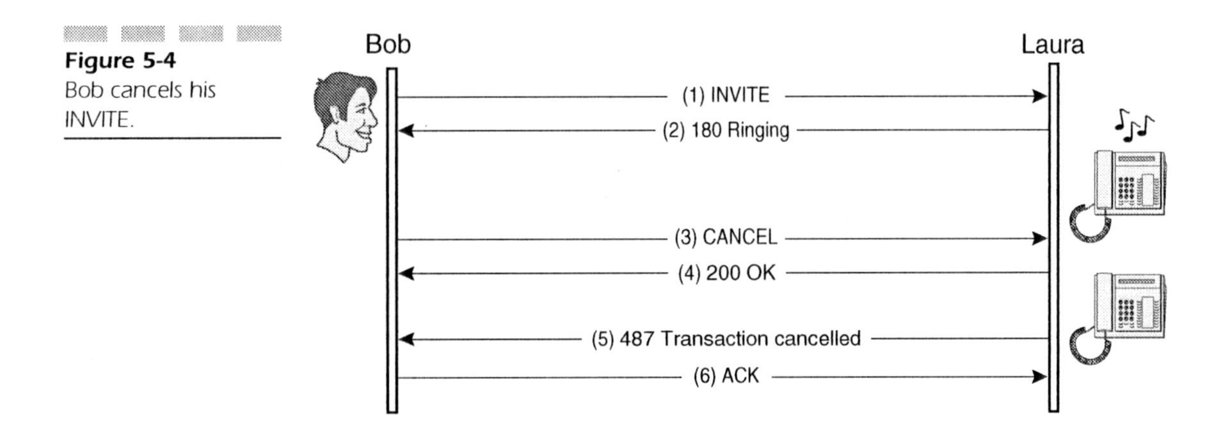

Figure 5-4 Bob cancels his INVITE.

where Bob might be reachable: SIP:Bob@131.160.1.112, SIP:Bob.Johnson@company.com, and SIP:Bob@university.com. When this proxy receives an INVITE from Laura to Bob, it will try these three locations in parallel (at the same time). The forking proxy sends three INVITEs, one to each location. Bob, who is currently working at 131.160.1.112, answers the call. The forking proxy receives a 200 OK from SIP:Bob@131.160.1.112 and it forwards this response to Laura's UA. Because the session is already established between Laura and Bob, the forking proxy wants to stop the other searches initiated, so it sends two CANCELs, one to each location, to close out the searches (Figure 5-5).

Remember that a CANCEL request does not affect a transaction once a final response has been sent. Therefore, in our example, even if the forking proxy sends a CANCEL to SIP:Bob@131.160.1.112, the session between Bob and Laura would persist. CANCEL cannot terminate an ongoing transaction. It is ignored by completed transactions.

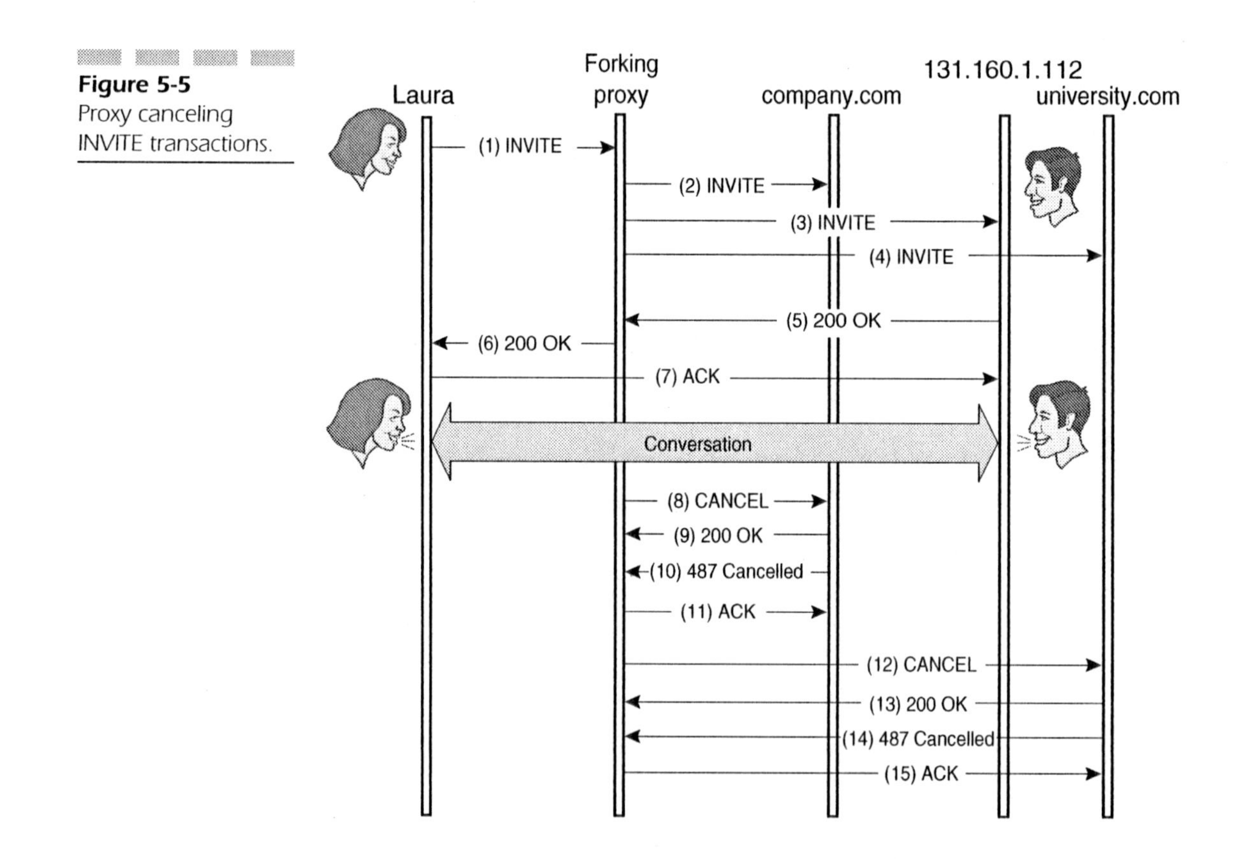

Figure 5-5 Proxy canceling INVITE transactions.

BYE BYE requests are used to abandon sessions. In two-party sessions, abandonment by one of the parties implies that the session is terminated. For instance, when Bob sends a BYE to Laura, their session is automatically terminated (Figure 5-6). In multicast scenarios, however, a BYE request from one of the participants just means that a particular participant leaves the conference. The session itself is not affected. In fact, it's common practice in large multicast sessions to not send a BYE when leaving the session.

REGISTER Users send REGISTER requests to inform a server (in this case, referred to as a registrar) about their current location. Bob can send a REGISTER to the registrar at company.com directing that all incoming requests for SIP:Bob.Johnson@company.com should be proxied, or redirected, to SIP:Bob@131.160.1.112 (Figure 5-7).

SIP servers are usually co-located with SIP registrars. A SIP registrar can send all information received in various REGISTER requests to a single location server, making it available to any SIP server trying to find a user.

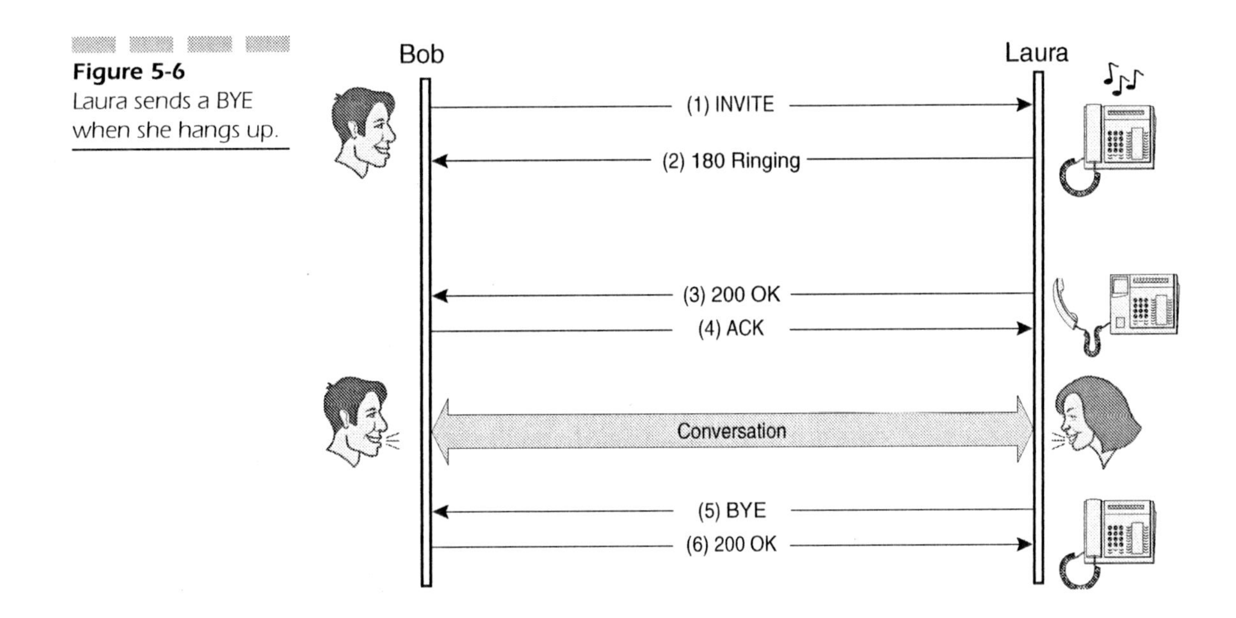

Figure 5-6 Laura sends a BYE when she hangs up.

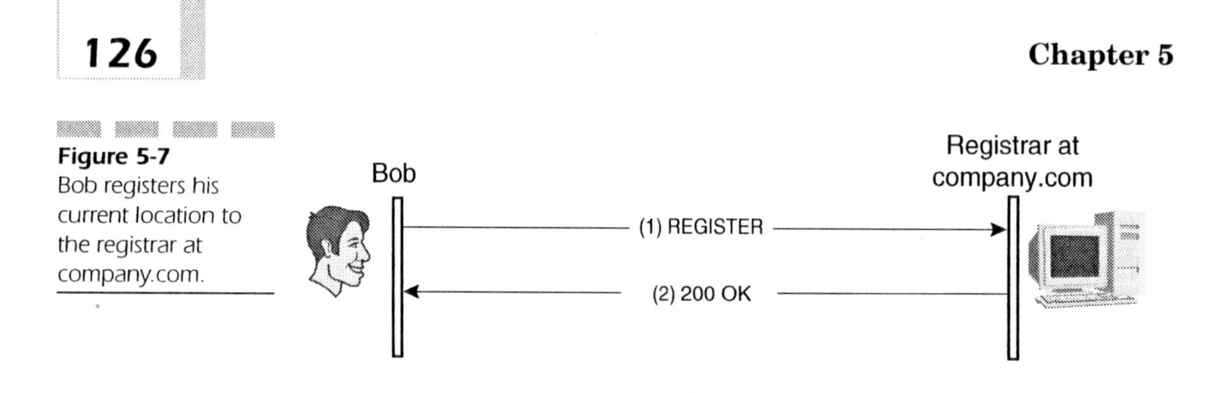

Figure 5-7 Bob registers his current location to the registrar at company.com.

REGISTER messages also contain the times when the registration pertains. For instance, Bob can register his present location until four o'clock in the afternoon because he knows that's when he will leave the office. A user can also be registered at several locations at the same time, indicating to the server that it should search for the user at all registered locations until he or she is reached.

OPTIONS OPTIONS requests query a server about its capabilities (Figure 5-8), including which methods and which session description protocols it supports. One SIP server might answer to an OPTIONS request that it supports SDP as session description protocol and five methods: INVITE, ACK, CANCEL, BYE, and OPTIONS. Because the server does not support the REGISTER method, I can deduce that it is not a registrar. The OPTIONS method might not look useful now, but as new extensions add new methods to SIP, the OPTIONS method is a great way to discover which methods a certain server supports.

An OPTIONS method also returns data that specifies which encodings for message bodies the server understands. If a certain server understands, for instance, a certain compression scheme, the client will be able to send the session descriptions compressed and take the opportunity to save some bandwidth.

Types of Proxy Servers

Proxy servers can be classified according to the amount of state information that they store during a session. SIP defines three types of proxy servers: call stateful, stateful, and stateless.

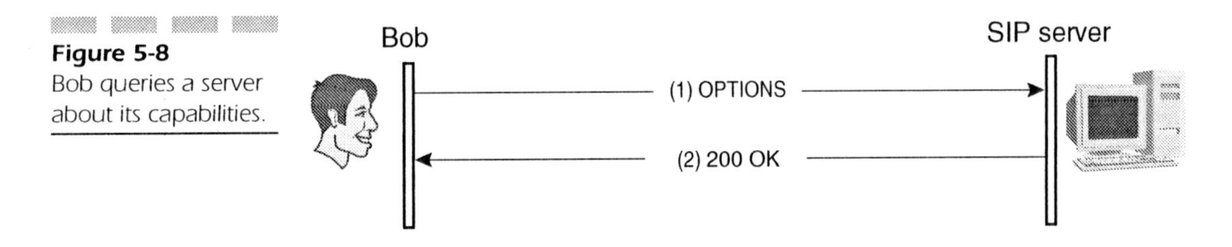

Figure 5-8 Bob queries a server about its capabilities.

Call Stateful Proxy

Call stateful proxies need to be informed of all the SIP transactions that occur during the session and therefore, they are always in the path taken by SIP messages traveling between end users. These proxies store state information from the moment the session is established until the moment it ends.

An example of a call stateful proxy is a server that implements a call-related service, such as receiving an e-mail at the conclusion of every call with information about the duration of each call (Figure 5-9). To calculate the length of the call, the proxy should be in the path of the INVITE that initiates the call and also in the path of the BYE that finishes the call.

Stateful Proxy

Stateful proxies are sometimes called transaction stateful proxies because the transaction is their sole concern. A stateful proxy stores state related to a given transaction until the transaction concludes. It does not need to be in the path taken by the SIP messages for subsequent transactions.

Forking proxies are good examples of stateful proxies (Figure 5-10). They send INVITEs to several different places and have to store state about the INVITE transaction in order to know whether all of the locations tried have returned a final response or not. However, once the user is reached at a particular location, the proxy does not need to remain in the signalling path any longer.

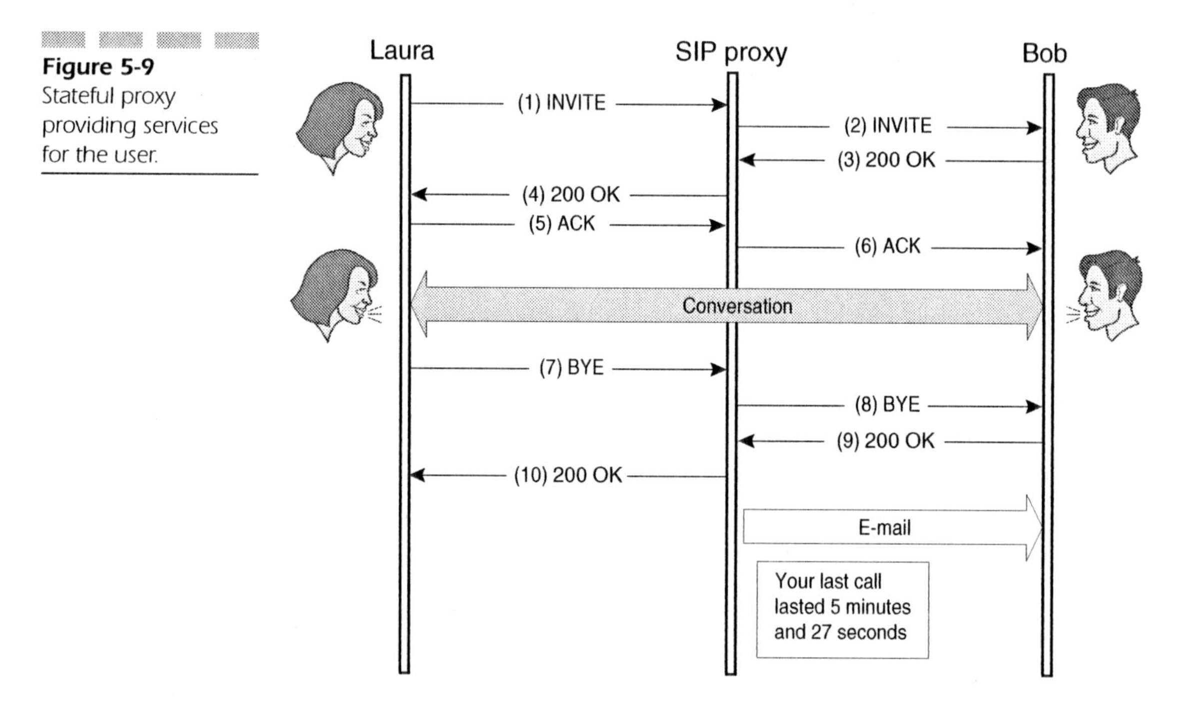

Figure 5-9 Stateful proxy providing services for the user.

Generation of ACKs Figure 5-10 also shows how ACKs are generated in SIP. We saw that ACKs are generated as final responses to an INVITE. They are part of the three-way handshake and, depending on the type of response the server returns, they are generated either by proxies or by the UAC.

Proxy servers can only ACK non-successful final responses, which have a status code that is greater than 299. Success responses (status code between 200 and 299) are always ACKed by the UA initiating the INVITE.

In Figure 5-10, the proxy server ACKs non-successful responses in messages (4) and (7). However, the UA ACKs the 200 OK response in message (11). This enables a proxy to try multiple locations without having to inform the originating UA about the unsuccessful attempts to locate the final user. Once a user responds positively to an INVITE, the originating UA must receive the remote session description in order to establish the session.

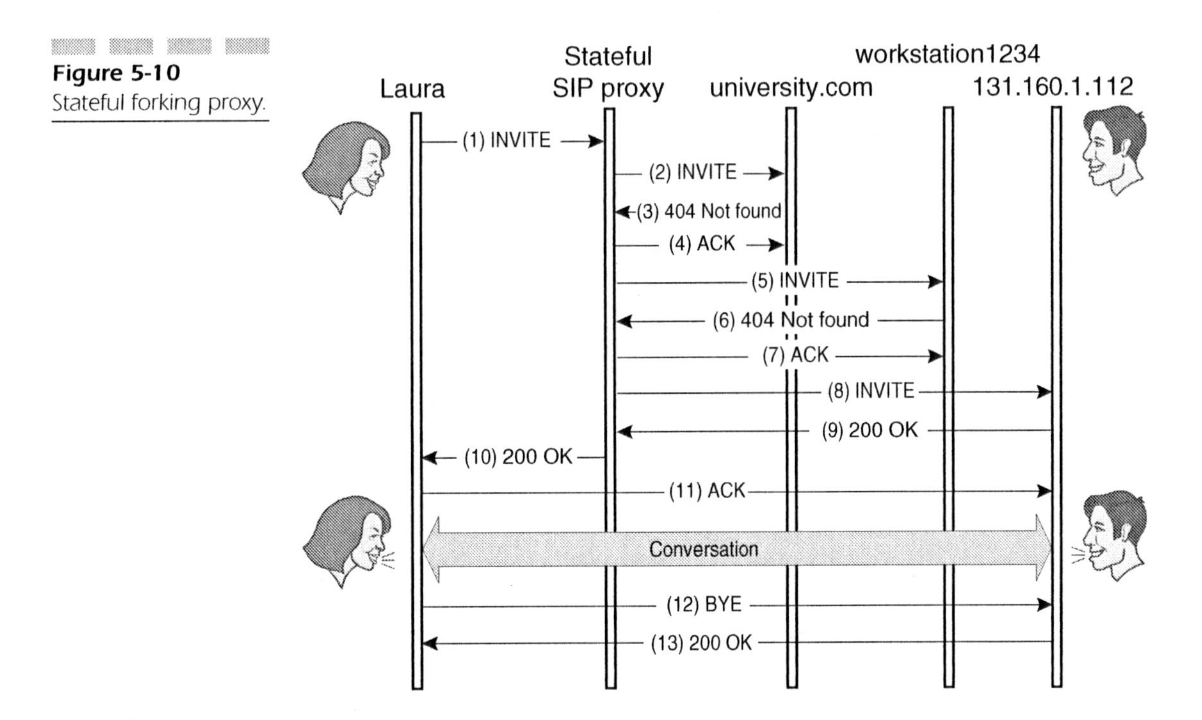

Figure 5-10 Stateful forking proxy.

Stateless Proxy

Stateless proxies don't keep any state. They receive a request, forward it to the next hop, and immediately delete all state related to that request. When a stateless proxy receives a response, it determines routing based solely on the Via header and it does not maintain state for it. (We'll discuss Via headers later in this chapter.)

Distribution of Proxies

An analysis of the IP traffic in a network invariably shows that the core is more stressed than the edges. This is true of SIP traffic as well.

SIP servers in the core need to be able to handle many messages, whereas SIP servers in the periphery do not have to support equally as heavy loads. SIP is designed for stateless servers at the core. They perform routing based on Request-URI or Via headers as fast and efficiently as we know how to do it. At the edges of the network, call stateful and stateful servers can be implemented to perform routing based on more complicated variables (such as the time of day or identity of the sender in the From field), or they can fork requests and provide services to the user.

Distributing servers this way makes SIP a very scalable protocol that can be implemented in increasingly large networks such as the Internet. SIP keeps the core fast and simple and pushes the intelligence to the periphery of the network (Figure 5-11).

Format of SIP Messages

Protocol design proceeds in discrete stages. When it has been decided which information will be exchanged between distributed systems, the next step is to decide how this information should be encoded. This decision has basically two approaches: binary, which uses bit fields to encode information,

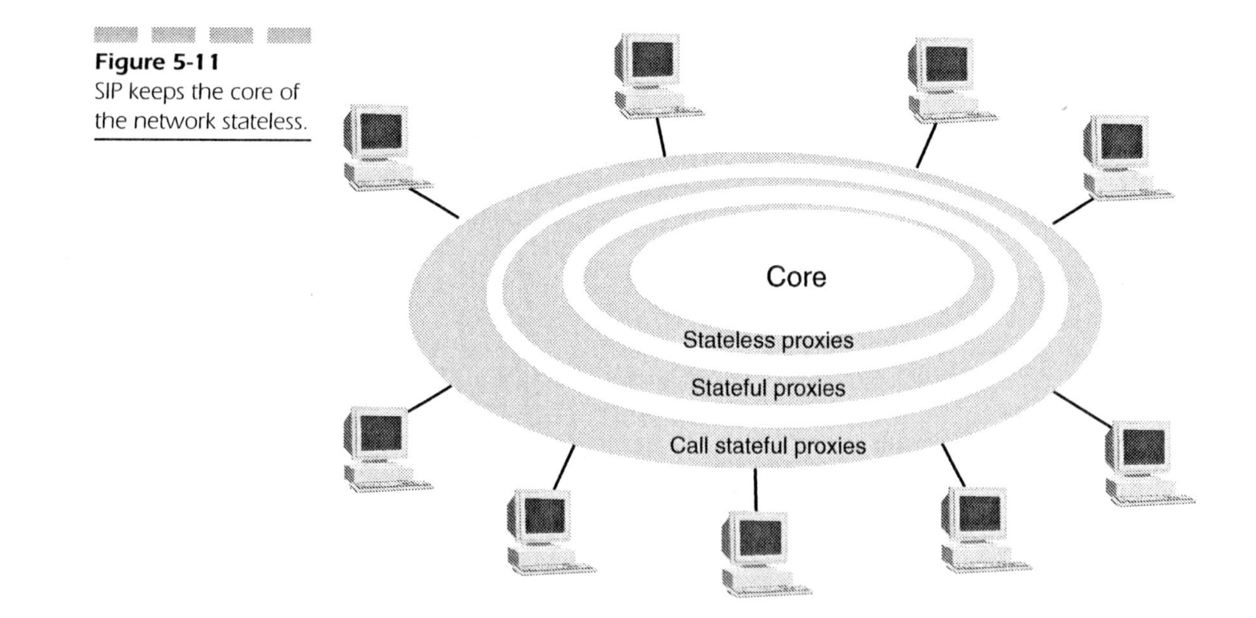

Figure 5-11 SIP keeps the core of the network stateless.

and textual, which uses strings of characters. The following example illustrates the differences between the two approaches.

Users need to keep track of the current month on their computers, and a server in the network has this information. We need a protocol that can transfer this information from server to desktop. The current month field can take exactly 12 possible values: January, February, March, April, May, June, July, August, September, October, November, and December.

A text-based protocol would transmit the name of the month between systems. Let's say that the message contents are January. Every character (letter) is typically encoded using one byte (8 bits). Thus, the message January will be encoded using 49 bits (7 letters times 8 bits).

A binary protocol, on the other hand, would define a table with possible values and their corresponding encoding, as shown in Table 5-3.

Thus, to transmit the current month, the binary protocol would send a 4-bit message containing 0000.

SIP uses text encoding as opposed to binary. This issue has created heated discussions. Text vs. binary seems to be a quasi-religious debate in which it is impossible to maintain a moderate opinion. Text proponents claim that text-based protocols are debugged more easily because they can be read directly by a human and that text protocols are more flexible and easier to extend with new features.

Binary believers argue that binary protocols use bandwidth more efficiently and can also be easy to debug and extend with the proper tools. Both types of encoding have advantages and disadvantages that we won't enumerate in this discussion, but keep in mind that SIP is a text-based protocol and exhibits all the pros and cons of text-based protocols in general.

Table 5-3

Binary Encoding of the Months

0000	January	0110	July
0001	February	0111	August
0010	March	1000	September
0011	April	1001	October
0100	May	1010	November
0101	June	1011	December

SIP Request Format

A SIP request consists of a request line, several headers, an empty line, and a message body. Table 5-4 shows the format of a SIP request. The message body is optional; some requests do not carry it.

Request Line A request line has three elements: method, Request-URI, and protocol version. The method indicates the type of request, and we've explained several in previous sections. The Request-URI indicates the next hop, which is where the request has to be routed. In Figure 5-12, the SIP proxy at company.com receives an INVITE with the Request-URI SIP:Bob.Johnson@company.com. This proxy knows that Bob might be reachable at two places so it generates two INVITEs. One will contain SIP:Bob@university.com as the Request-URI and it will be sent to the server at university.com. The second INVITE will have SIP:Bob@131.160.1.112 and will be sent to 131.160.1.112. Hence, the Request-URI contains the address of the next hop in the path.

Finally, we know the protocol version to be SIP/2.0. Therefore, the request line of the INVITE received in the previous example by the server at company.com would look like the following:

```
INVITE sip:Bob.Johnson@company.com SIP/2.0
```

SIP Response Format

A SIP response consists of a status line, several headers, an empty line, and a message body. Table 5-5 shows the format of a SIP response. The message body is optional; some responses do not carry it.

Table 5-4 Format of an SIP Request

Request-line
Several headers
Empty line
Message body

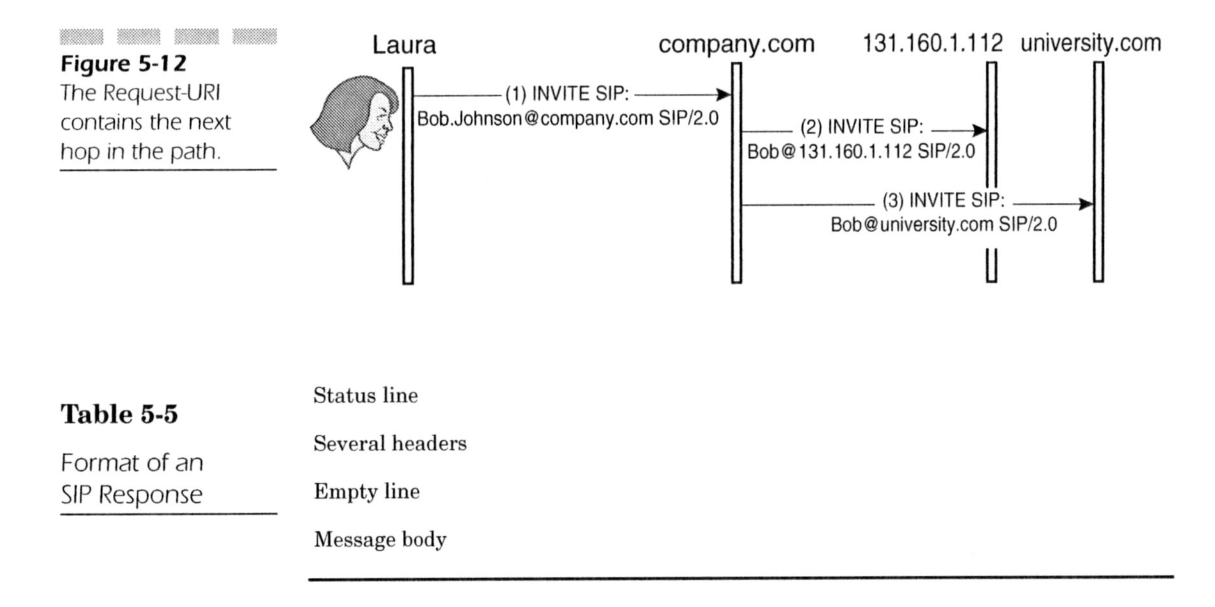

Figure 5-12 The Request-URI contains the next hop in the path.

Table 5-5 Format of an SIP Response

Format
Status line
Several headers
Empty line
Message body

Status Line A status line has three elements: protocol version, status code, and a reason phrase. The current protocol version is written as SIP/2.0. The status code reports transaction status. As described earlier, status codes are integers from 100 to 699 and are grouped into six different classes (refer to Table 5-1). The reason phrase is meant for human eyes only. It is not meaningful for computers processing SIP responses. Below there is an example of a status line.

```
SIP/2.0 180 Ringing
```

Reliable Transmission of Responses Final responses are transmitted reliably between server and client, using retransmissions or a reliable transport protocol to ensure delivery. Provisional responses are not. They may either be received by the client or be lost in the network. SIP takes this approach because it is more concerned with whether a session was established or not, and the reasons why it wasn't, than with how the session setup is progressing.

Ergo in a SIP call, for instance, callers are guaranteed notification that the call has been accepted, but might not know when the callee alert began (Figure 5-13). SIP can be extended for reliable delivery of provisional responses if necessary.

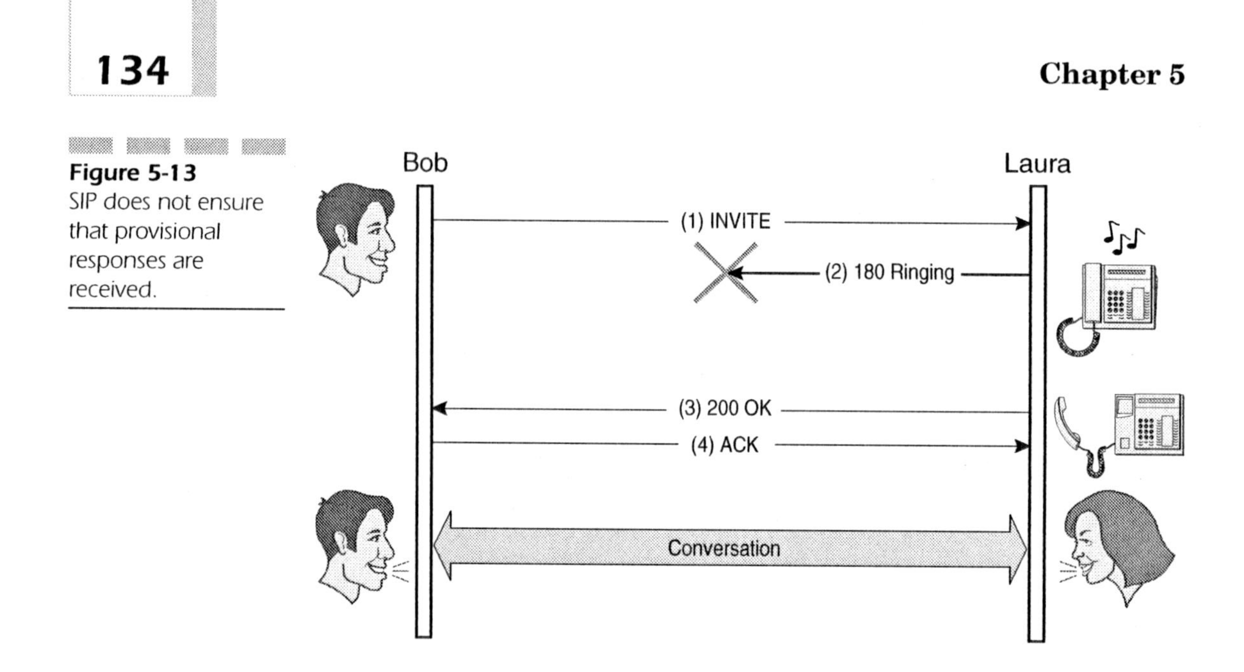

Figure 5-13 SIP does not ensure that provisional responses are received.

SIP Headers

SIP requests contain some SIP headers after the request line, whereas SIP responses put them after the status line. Headers provide information about the request (or response) and about the body it contains. Some headers can be used in both requests and responses, but others are specific to requests (or responses) alone. The header consists of the header name, followed by a colon, followed by the header value.

For instance, the header called From, which identifies the originator of a particular request, looks like the following:

```
From: Bob Johnson <sip:Bob.Johnson@company.com>
```

As can be seen in this example, a header value can have several fields. In this example, the From header has two fields: a person's name and his SIP URL.

Table 5-6 contains the SIP headers defined in the core protocol.

In the following sections, I'll explain the purpose of the most important SIP headers and give simple usage examples of them.

Table 5-6
SIP Headers

Accept	Content-encoding	Max-forwards	Route
Accept-encoding	Content-language	MIME-version	Server
Accept-language	Content-length	Organization	Subject
Alert-info	Content-type	Priority	Supported
Allow	Cseq	Proxy-authenticate	Timestamp
Also	Date	Proxy-authorization	To
Authorization	Encryption	Proxy-require	Unsupported
Call-ID	Error-info	Record-route	User-agent
Call-info	Expires	Require	Via
Contact	From	Response-key	Warning
Content-disposition	In-reply-to	Retry-after	WWW-authenticate

Call-ID The Call-ID represents a SIP signalling relationship shared among two or more users. It identifies a particular invitation and all of the subsequent transactions related to that invitation in a format that looks like the following:

```
Call-ID: ges456fcdw21lkfgte12ax@workstation1234.university.com
```

A server that is juggling SIP signalling for many sessions employs Call-ID to associate incoming messages to the proper session. For instance, Bob invites Laura to a chess session with a particular Call-ID. Laura's UA accepts and soon the game commences. After a while, Bob calls Laura to speak with her while they are still playing chess. This INVITE from Bob's UA has a different Call-ID from the previous one.

When Bob and Laura finish speaking, Bob's UA sends a BYE to Laura's UA to end the phone call. Laura's UA uses the Call-ID of the BYE message to decide whether to terminate the chess game or the conversation (Figure 5-14).

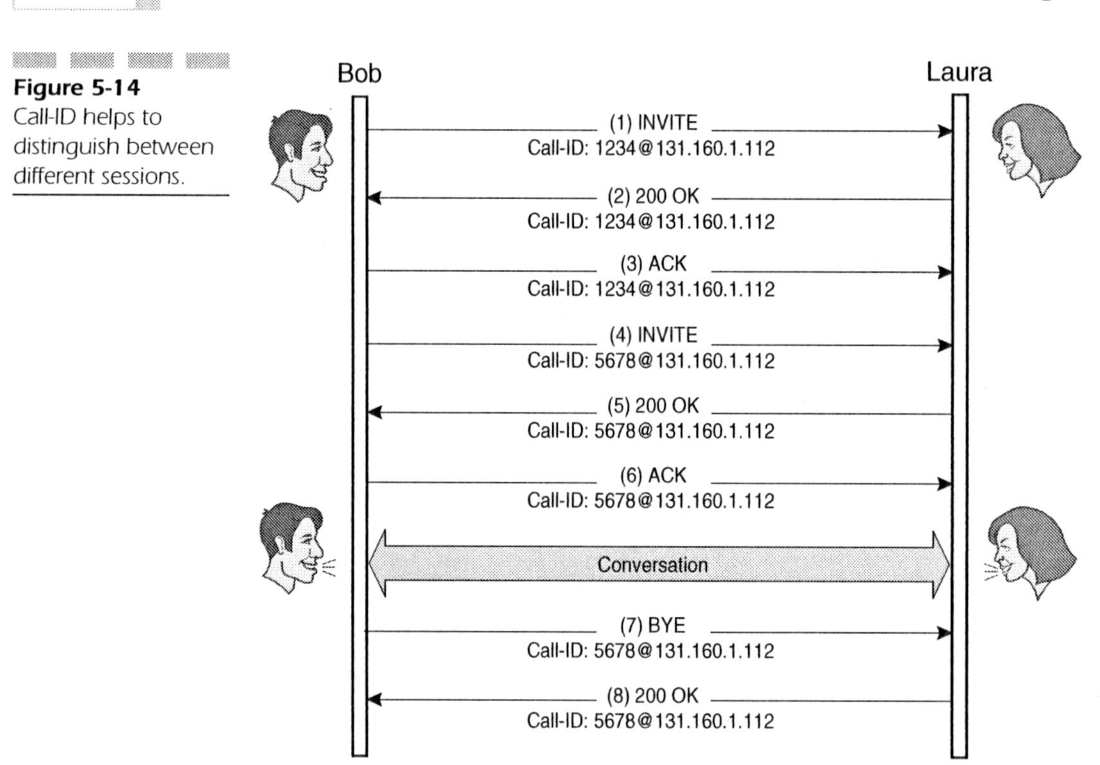

Figure 5-14 Call-ID helps to distinguish between different sessions.

Contact A Contact header provides a URL where the user can be reached directly. This feature is important because it offloads SIP servers that do not need to be in the signalling path after routing the first INVITE.

For instance, Laura calls Bob at SIP:Bob.Johnson@company.com. Company.com's proxy forwards the INVITE to SIP:Bob@131.160.1.112, where Bob turns out to be. He accepts the call. Bob's UA returns a 200 OK response with a Contact header:

```
Contact: Bob Johnson <sip:Bob@131.160.1.112>
```

When Laura's UA receives this 200 OK response, it sends the ACK to Bob's UA. Because Bob's location can be found in the contact header, the ACK is sent directly to SIP:Bob@131.160.1.112 and the ACK does not traverse the proxy at company.com.

Figure 5-15 shows how subsequent requests, such as the BYE, are sent directly between session participants.

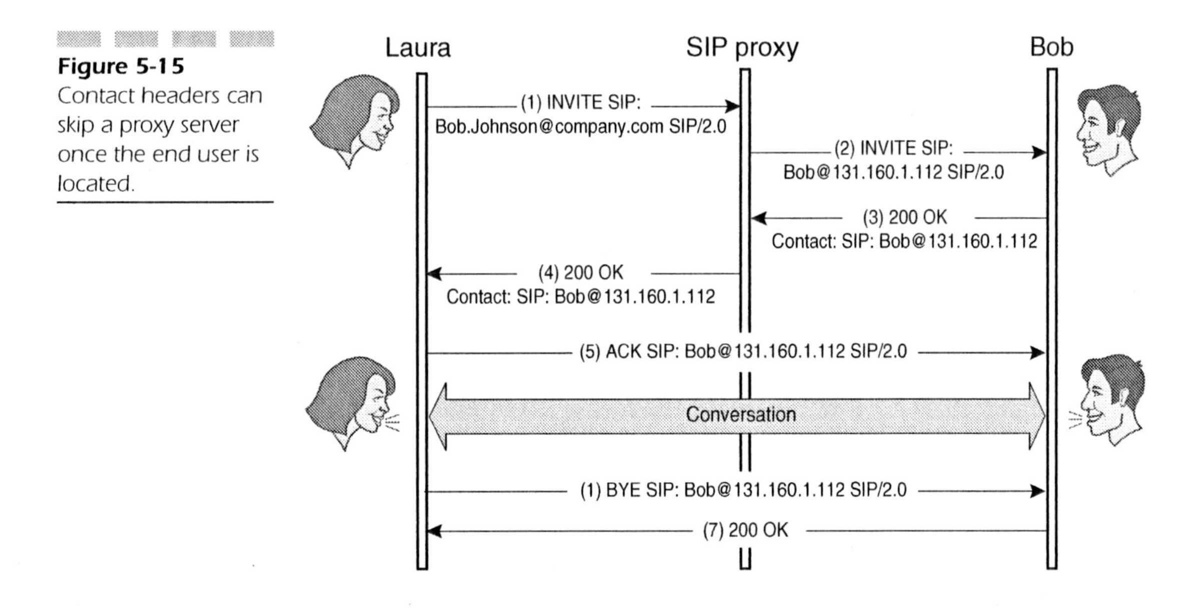

Figure 5-15 Contact headers can skip a proxy server once the end user is located.

Cseq The *Command Sequence* (Cseq) header has two fields: an integer and a method name. The numerical part of the Cseq is used to order different requests within the same session (defined by a particular Call-ID). It is also used to match requests against responses. For instance, Bob sends an INVITE to Laura with the following Cseq:

```
Cseq: 1 INVITE
```

Laura returns a 200 OK response with the same Cseq as the INVITE. If Bob wants to modify the session already established, he will send a second INVITE (re-INVITE) with the following Cseq:

```
Cseq: 2 INVITE
```

If a retransmission of the 200 OK response is delayed by the network and arrives at Bob's UA after it has generated the second INVITE, it knows that this was a response for the first INVITE, thanks to the Cseq header (Figure 5-16).

After an INVITE all subsequent requests (except ACK and CANCEL) contain a Cseq which is the result of incrementing by one of the Cseqs of the original request.

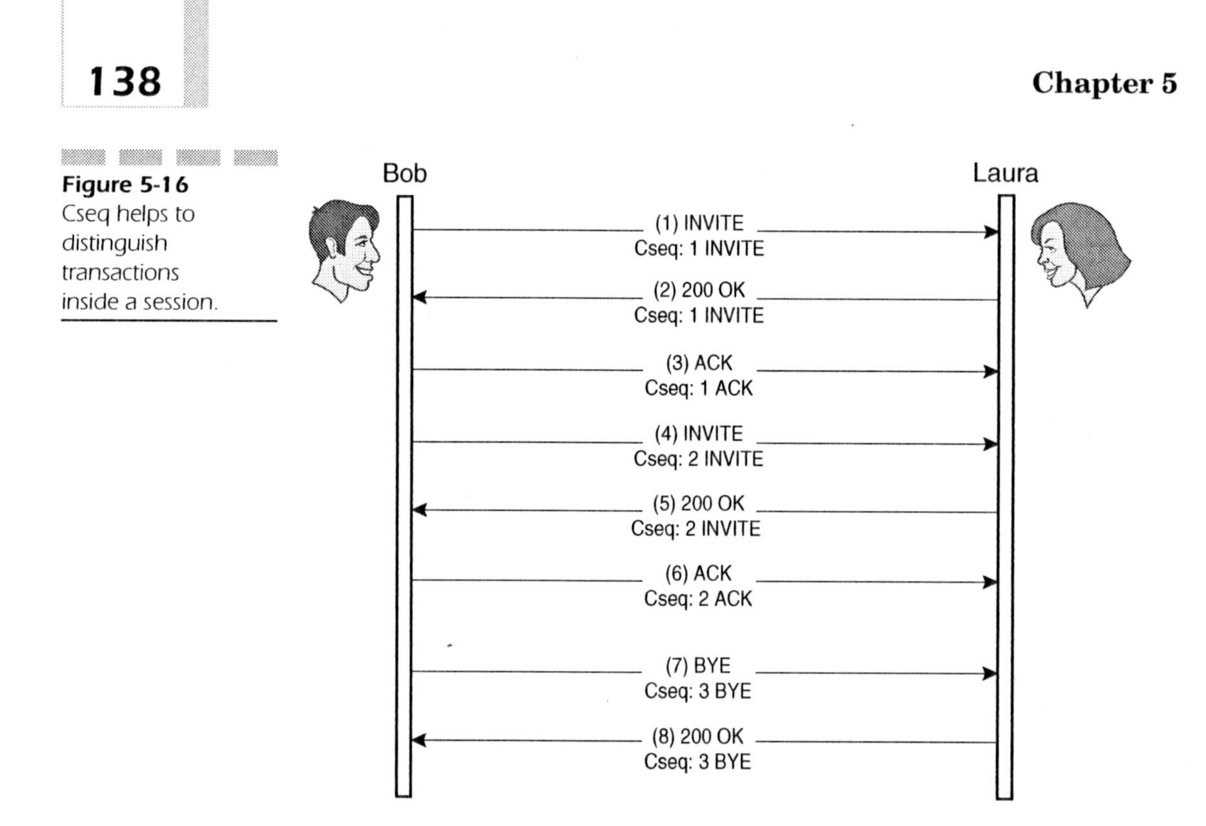

Figure 5-16 Cseq helps to distinguish transactions inside a session.

Cseq in ACK An ACK request has the same Cseq as the INVITE it acknowledges. This enables proxies to generate ACKs for non-successful final responses without creating new Cseqs. In fact, new Cseqs can only be created by the UA, which ensures that Cseqs are unique.

Cseq in CANCEL A CANCEL request has the same Cseq as the request it cancels. This also enables proxies to generate CANCELs without creating new Cseqs. Moreover, CANCEL is the reason why the Cseq header includes a method name after the numerical part.

Because the Cseq number of the INVITE and the CANCEL is the same, a SIP client couldn't distinguish responses for CANCEL and responses for INVITE without an additional field. The method name inside Cseq solves the problem (Figure 5-17).

From The From header contains the initiator of the request and a SIP URL:

```
From: Bob Johnson <sip:Bob.Johnson@company.com>
```

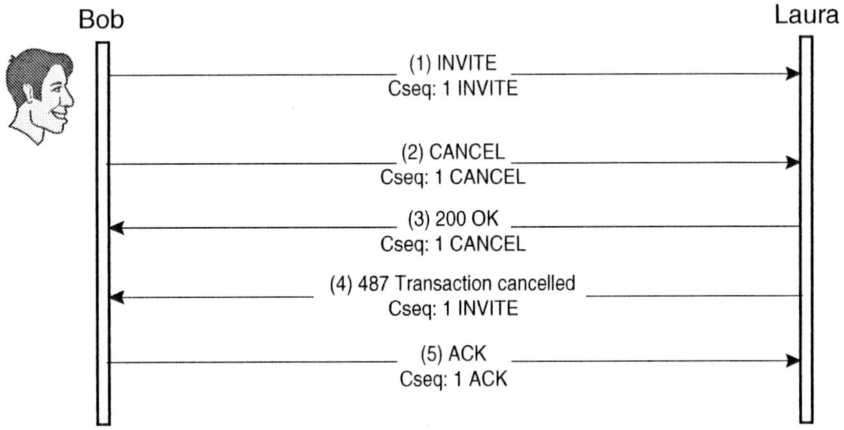

Figure 5-17 The method name in the Cseq permits differentiating responses for INVITEs and CANCELs.

Bob — Laura

(1) INVITE
Cseq: 1 INVITE

(2) CANCEL
Cseq: 1 CANCEL

(3) 200 OK
Cseq: 1 CANCEL

(4) 487 Transaction cancelled
Cseq: 1 INVITE

(5) ACK
Cseq: 1 ACK

Record-Route and Route These two headers are used by proxies that want to be in the signalling path for the entire session. We saw that Contact headers enable the UAs to send requests directly to each other. This creates offloading proxies in the path; they route the first INVITE to the proper destination and then let the UAs begin to exchange SIP signalling. However, sometimes a proxy needs to stay in the signalling path, in which case a mechanism is needed to keep UAs from exchanging SIP messages on their own. This mechanism consists of two headers: Route and Record-Route.

A proxy may want to remain in the signalling path after the first INVITE for many reasons. One of them is security. Some domains have a security proxy, a firewall, that filters incoming SIP messages. SIP messages that do not successfully traverse the security proxy are not accepted into the domain. Another reason is service provision. A proxy that provides a session-related service needs to know as a matter of course when the session is over; for our purposes, this is when one UA sends a BYE request to another. We saw already an example of such a service in Figure 5-9. The call stateful proxy of that example had to see the BYE from Laura to Bob in order to e-mail Bob with information about the duration of the call.

Figure 5-18 illustrates how these two headers work. Laura sends an INVITE to Bob. The INVITE traverses a SIP proxy that wishes to be in the signalling path for subsequent requests between Laura and Bob. The proxy adds a Record-Route header containing its address to the INVITE. Bob's

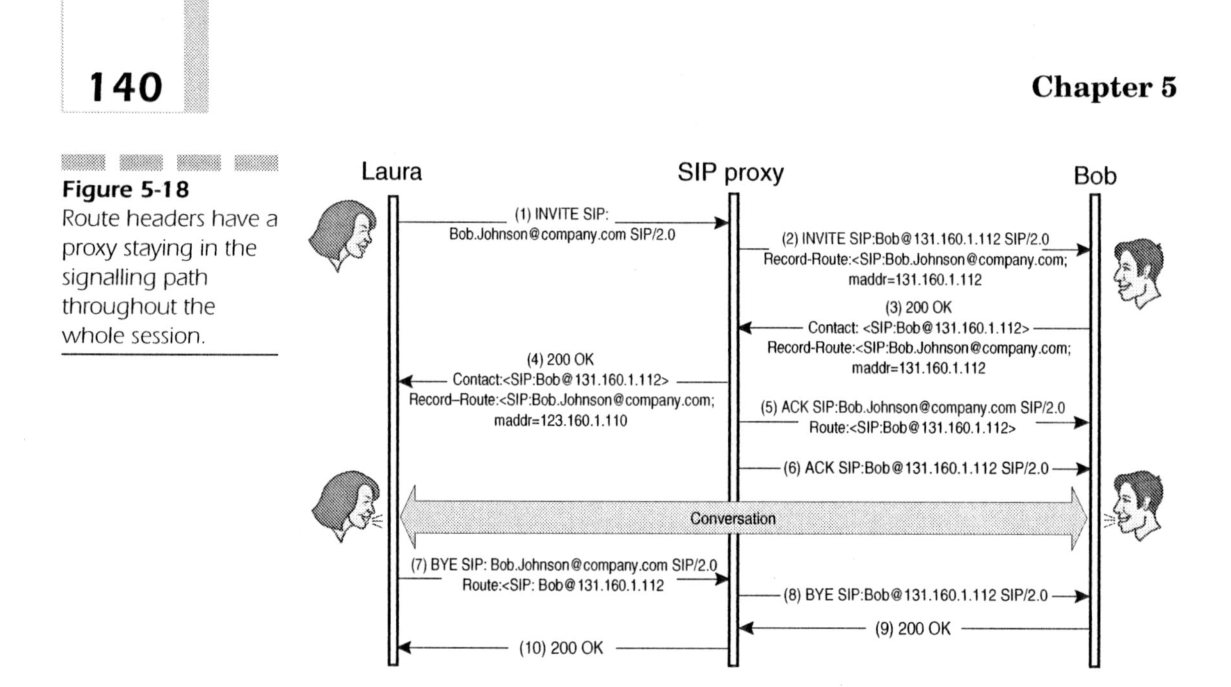

Figure 5-18 Route headers have a proxy staying in the signalling path throughout the whole session.

UA receives the INVITE complete with this Record-Route header and includes it in the 200 OK response. Bob's UA also adds its Contact header to the response.

The maddr parameter that appears in the Record-Route just contains the IP address of the server, which is added to record the server's real IP address for future requests.

Laura's UA receives the 200 OK response and builds a Route header that will be used in subsequent requests. The Route header is built from both the Record-Route and the Contact header present in the response. Because only one proxy needs to be in the signalling path, all the subsequent requests from Laura to Bob (ACK and BYE in this example) will be sent to it and will contain a Route header with Bob's Contact address. This way, the proxy knows to send the request to the address contained in the Route header.

Several Proxies The previous example shows how Record-Route works to inform Laura's UA that subsequent requests have to be sent through the proxy rather than directly to Bob. However, it does not show why the Route header is needed. A scenario with more proxies can help explain the purpose of the Route. Figure 5-19 contains three proxies: P1, P2, and P3. P1 and P3 have to be in the signalling path, but P2 does not. We can see how the ACK from Laura to Bob contains a Route header telling P1 to forward the request to P3. The last address in the Route header is Bob's Contact address.

Note that Figure 5-19 does not contain the real format of Contact, Record-Route, and Route headers as Figure 5-18 did. Instead, it uses a symbolic format that just indicates which addresses are contained in each header.

To The To header always contains the recipient of the request. It usually contains the public address of the destination party as well. It is important to make a distinction between the To header of a request and the Request-URI. The To header, which remains the same throughout the session, is intended for the remote UA. It cannot be changed by proxies.

The Request-URI contains the address of the next hop in the signalling path and is therefore changed by every proxy in the path. Figure 5-20 illustrates the use of each.

Laura calls Bob using his public address, SIP:Bob.Johnson@company.com. This SIP URL will be inserted in the To header and it will not vary during the session; that is, all of the requests from Laura to Bob will have the same To field.

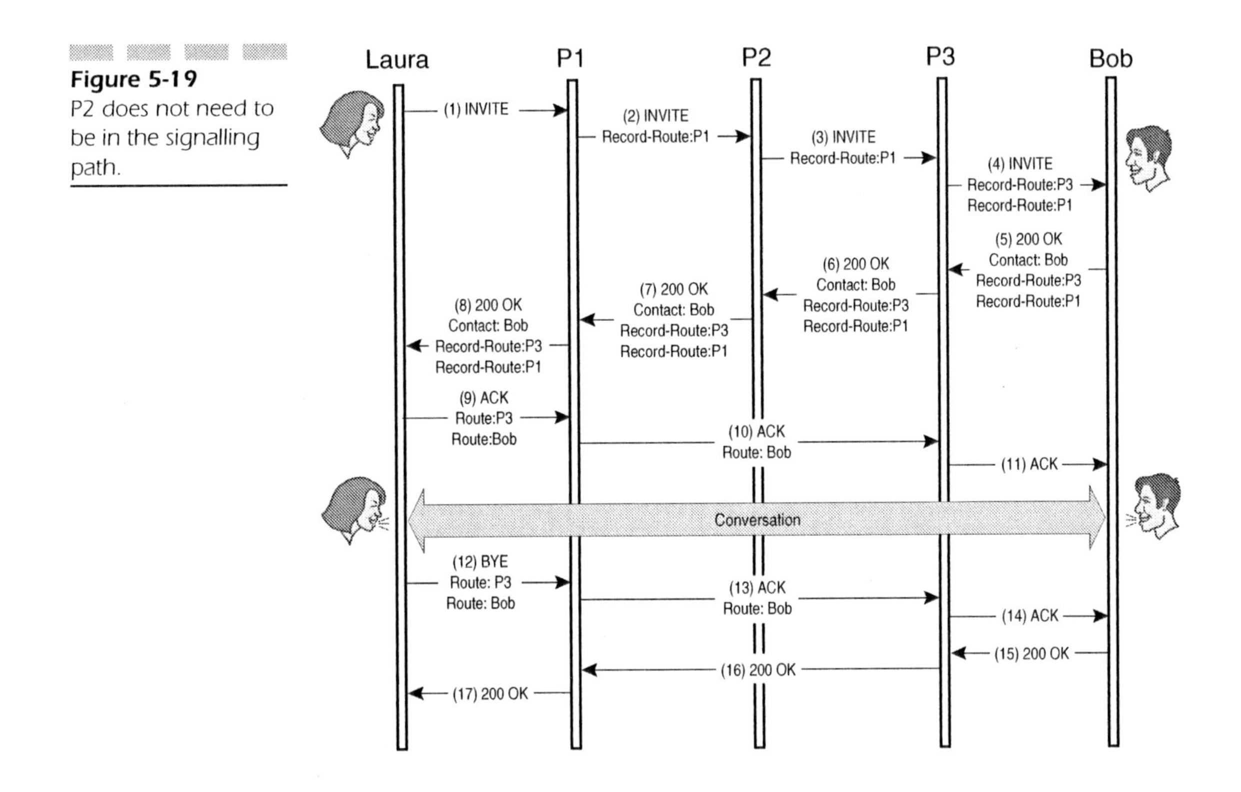

Figure 5-19
P2 does not need to be in the signalling path.

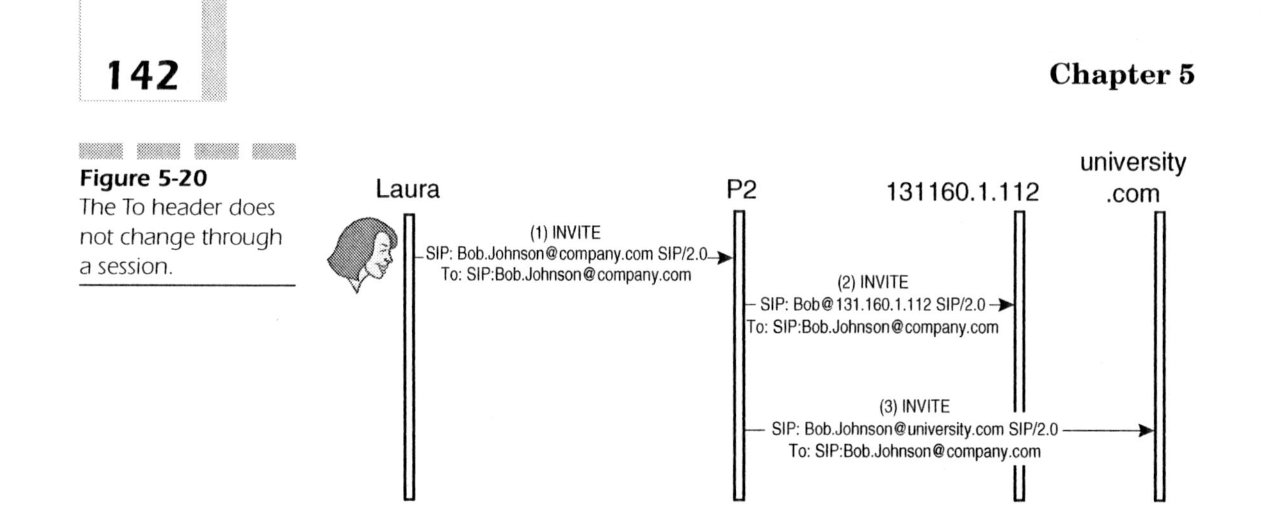

Figure 5-20 The To header does not change through a session.

Laura places the same SIP URL in the Request-URI so it will send the request to the SIP at company.com. This proxy performs a parallel search by trying two different SIP URLs: SIP:Bob@131.160.1.112 and SIP:Bob@ university.com.

Both INVITEs sent by the proxy at company.com have the same To header, but they have different Request-URIs.

Via Via headers store all the proxies that handle the request. Hence, they contain the path taken by the request (Figure 5-21). This information is used for detecting routing loops. If a request is forwarded in a loop, any proxy can notice it simply by inspecting the Via headers. If it finds its address there, the proxy knows that it has already handled this request. A typical Via header looks like the following:

```
Via: SIP/2.0/UDP workstation1234.company.com
```

Via headers are also used to route responses towards the client who generated the request. This way, a SIP response traverses the same set of proxies as the request, but in the opposite direction.

SIP Bodies

Both requests and responses may contain message bodies, separated from the message headers by an empty line. The message body carried by SIP messages is usually a session description, but it can consist of any opaque object. Because SIP proxies do not need to examine the message

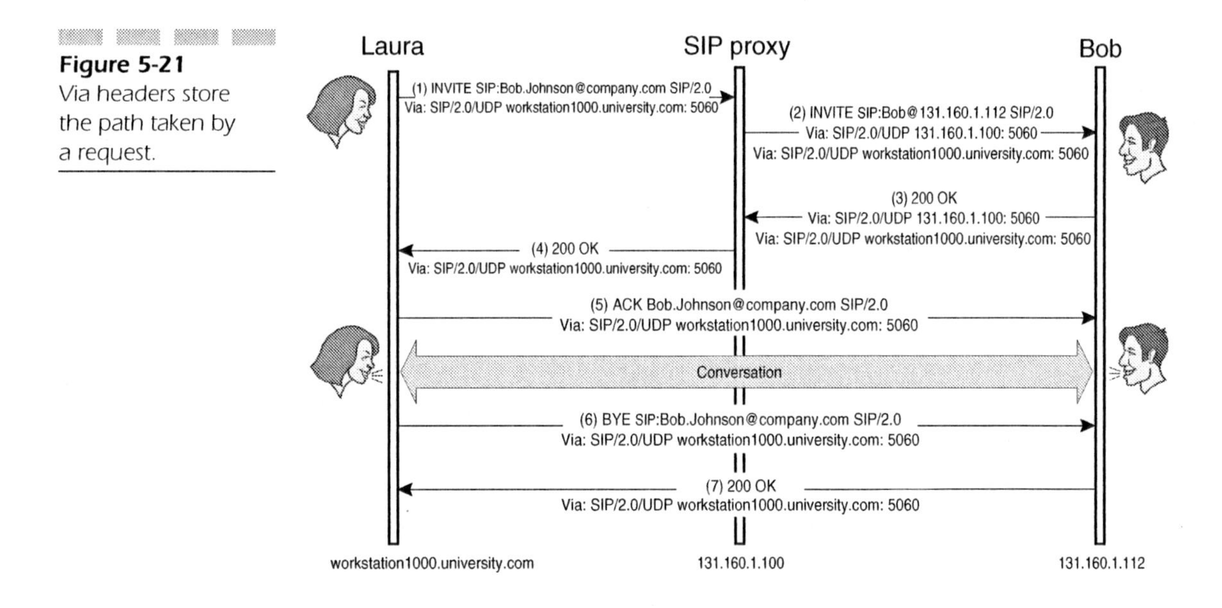

Figure 5-21 Via headers store the path taken by a request.

body, content is transparent to them. As a result, session descriptions are transmitted end to end between UAs. All information proxies need in order to route SIP messages is contained in the request and status lines and in the SIP headers. Because SIP bodies are only meaningful to the UA, message bodies can be encrypted end to end without losing any functionality.

Some proxies, however, might want to examine the session description. An example is a security proxy (firewall) that wants information about media being exchanged so that it can exclude unauthorized flows. For instance, if a company decides that its employees cannot establish videoconferences, the firewall can intercept all video streams while still letting audio streams pass through.

The following is an example of an SDP session description in a SIP body:

```
v=0
o=Bob 2890844526 2890842807 IN IP4 131.160.1.112
s=I want to know how you are doing
c=IN IP4 131.160.1.112
t=0 0
m=audio 49170 RTP/AVP 0
```

Just as e-mail messages can carry more than one attachment, SIP messages can carry several bodies. For instance, Laura may send an INVITE with two bodies: a session description and her photo. That way, Bob's UA can display her photo on the screen while Bob is alerted.

Transport Layer

We learned previously that SIP is an application layer protocol. Therefore, it makes use of transport layer protocols to transmit requests and responses. The behavior of any application layer protocol varies with the type of transport used. If it's reliable, the application layer protocol builds a message and delivers it to the transport layer, fully expecting that the message will arrive at the destination. The application layer does not know how the transport layer accomplishes its delivery; it just knows that the task is performed.

How is it performed? Typically, the transport layer will retransmit the message until the other end receives it and sends back some type of acknowledgement message. These retransmissions are transparent to the application layer.

On the other hand, if an application layer protocol makes runs on top of an unreliable transport layer protocol such as UDP, it cannot assume delivery. Therefore, application layer retransmissions have to be implemented. They are typically implemented as follows.

The application layer protocol builds a message and passes it to the transport layer. Should it fail to receive a confirmation of reception from the destination in a certain period of time, it will build the same message again and pass it to the transport again.

By utilizing application layer time-outs with its retransmissions, an application layer protocol can still exploit unreliable transport mechanisms.

Let us now see how SIP works over both types of transports.

INVITE Transactions

Because INVITE transactions involve a three-way handshake and an ACK request, they require different handling than any other transaction. Therefore, SIP entities treat INVITEs and ACKs in different ways than other methods.

Hop-by-Hop Treatment Remember that when proxies are in the path between two UAs, different transport protocols may also be between them (Figure 5-22). A UA using a reliable transport protocol towards a proxy cannot assume that same transport will be used end to end until the remote UA is reached.

SIP provides a mechanism to ensure that the INVITE will eventually be delivered; namely, it makes proxies responsible for getting an INVITE to the next hop in its path. Note that stateless proxies cannot assume this responsibility because they don't maintain the state information needed for retransmission when an INVITE gets lost. Therefore, next hop with respect to transport refers to the next stateful proxy (or the destination UA).

Transmitting an INVITE Because both a UA and a proxy have the same responsibility of ensuring that the INVITE reaches the next hop, the mechanisms used between a UA and a proxy, between two proxies, and between a proxy and a UA are exactly the same. In this section, we will explain the behavior of a UA sending an INVITE to a proxy, but that proxy will use exactly the same mechanisms towards the next proxy in the path.

A SIP UA sending an INVITE to a proxy over a reliable transport protocol does not have to perform any special task, but if an unreliable transport protocol such as UDP is used, it must be prepared to retransmit, sometimes repeatedly, until a previrional response is received. (Figure 5-23).

Proxy servers receiving an INVITE always generate a 100 Trying provisional response. A UA receiving an INVITE can generate any provisional response, such as 180 Ringing.

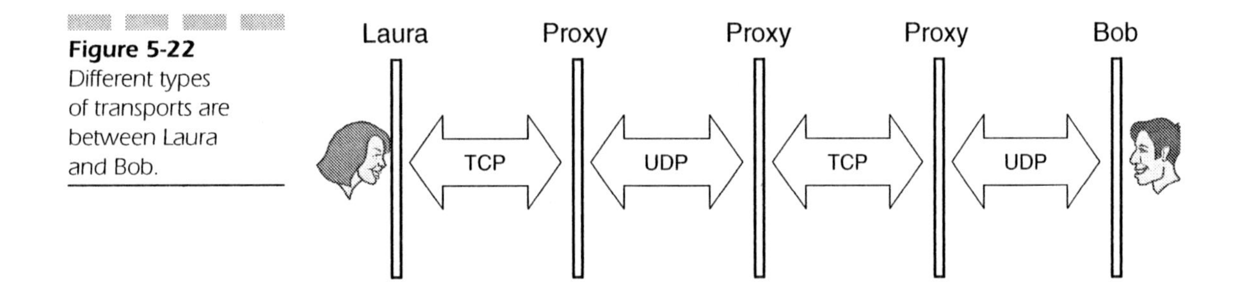

Figure 5-22 Different types of transports are between Laura and Bob.

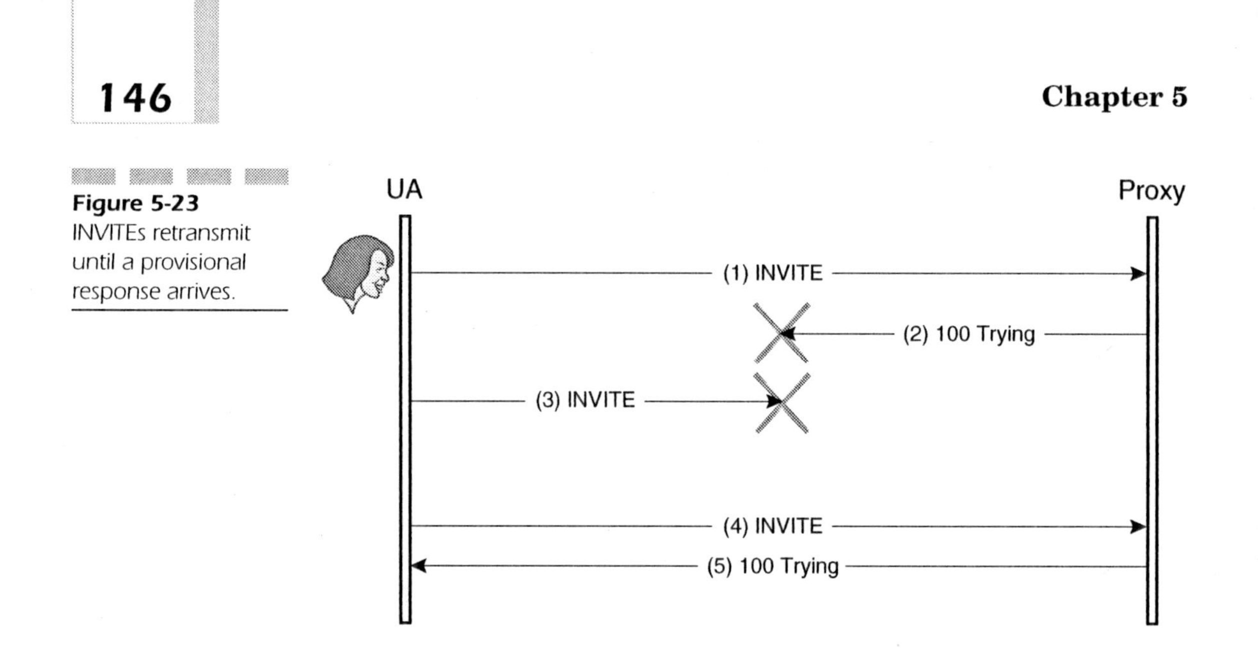

Figure 5-23 INVITEs retransmit until a provisional response arrives.

Transmitting Responses to an INVITE We have seen that provisional responses are used to prevent hop-by-hop INVITE retransmissions. However, nothing ensures that a provisional response from the callee's UA will reach the caller's UA. Proxies in the path will usually forward the response to the previous hop once, but will not retransmit if it fails to arrive (Figure 5-24).

In contrast, SIP guarantees that final responses arrive at their intended destination. Successful responses (200 to 299) are delivered reliably to the originating UA. Non-successful final responses (300 to 699) use the same hop-by-hop mechanism as INVITE.

Non-successful Final Responses The idea behind transmitting non-successful final responses is the same as that behind transmitting INVITEs. Every server ensures that the previous hop receives the response and then the previous hop assumes responsibility for handling the response.

A UA using an unreliable transport protocol is going to retransmit the non-successful final response until an ACK arrives (Figure 5-25).

In theory, a UA using a reliable transport protocol would not have to make use of the ACK. However, in order to make the protocol look homogeneous across reliable and unreliable transports, ACKs are used in both.

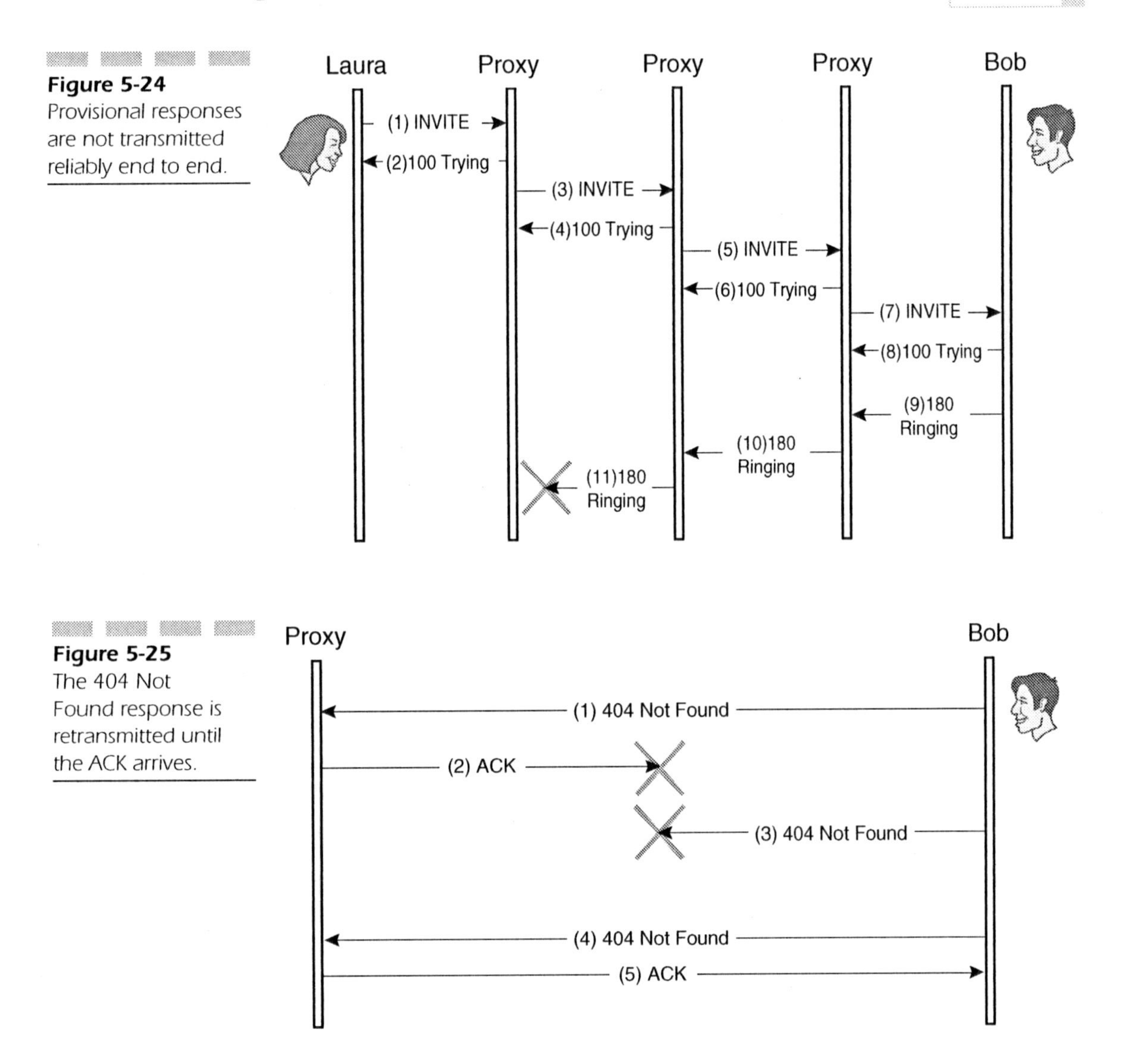

Figure 5-24 Provisional responses are not transmitted reliably end to end.

Figure 5-25 The 404 Not Found response is retransmitted until the ACK arrives.

I've mentioned some situations in which non-successful final responses are not transmitted to the originating UA. Figure 5-26 shows how the forking proxy at company.com receives 404 Not Found responses that it does not forward to Laura's UA, thanks to the hop-by-hop transport mechanism used for this kind of response.

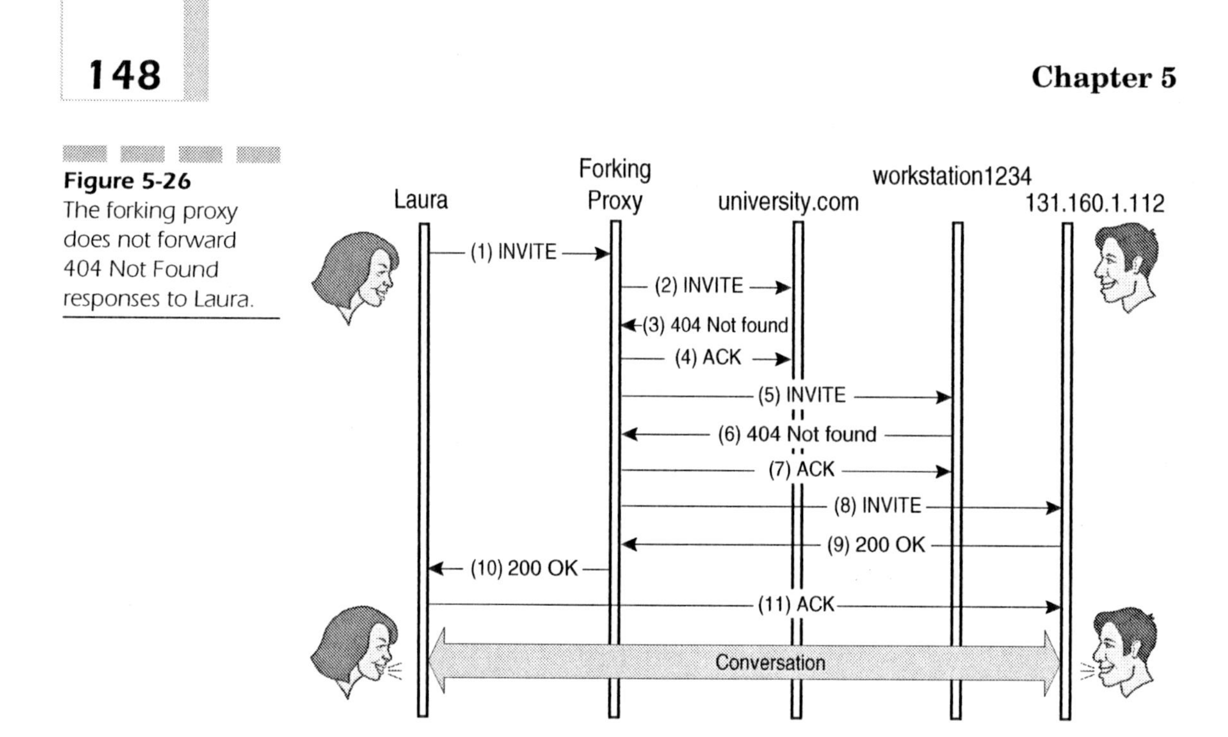

Figure 5-26 The forking proxy does not forward 404 Not Found responses to Laura.

Successful Final Responses Successful final responses are transmitted reliably end to end between UAs and do not need the hop-by-hop mechanism used by other final responses (Figure 5-27). Only the UA that originated an INVITE can send an ACK for a final successful response. Therefore, regardless of the transport protocol used (reliable or unreliable), a UA retransmits successful final responses until it receives an ACK from the originating UA.

Proxies in the path simply forward successful final responses and their ACKs. They are not implicated in reliability.

Figure 5-28 shows the whole session establishment process from the INVITE until the actual conversation takes place.

CANCEL Transactions

As hop-by-hop transactions, CANCEL transactions are handled in a special way. When a UA sends a CANCEL to a proxy, the proxy responds with a final response. At that point, the CANCEL transaction is finished for the UA. Next, the proxy will send another CANCEL to the next hop, and it will also receive a final response. You can see that reliability for CANCEL requests is easy to accomplish by retransmission (Figure 5-29).

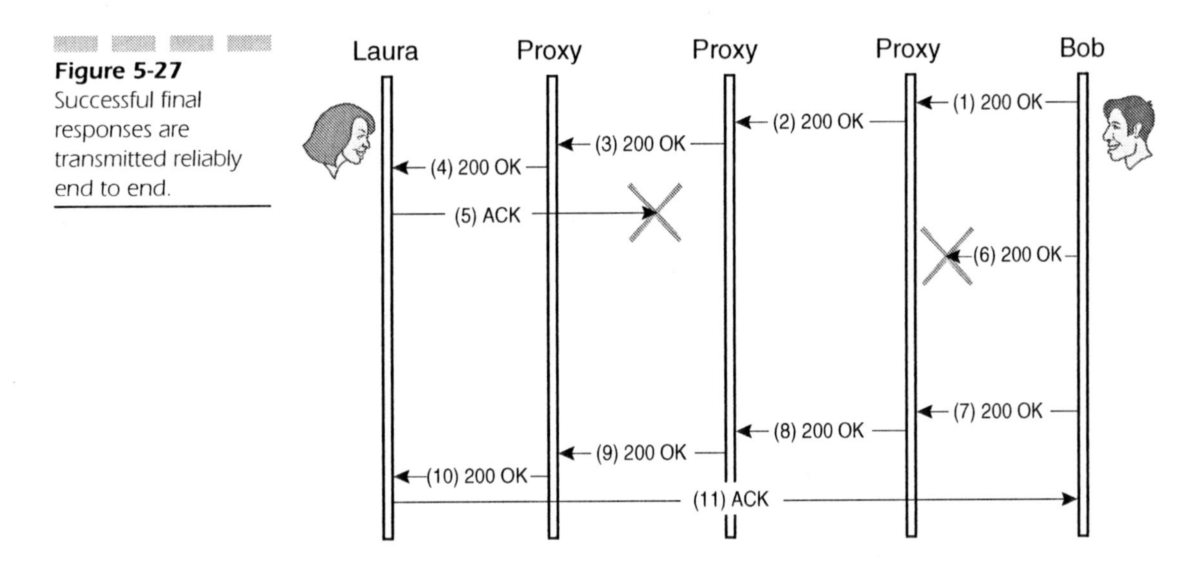

Figure 5-27 Successful final responses are transmitted reliably end to end.

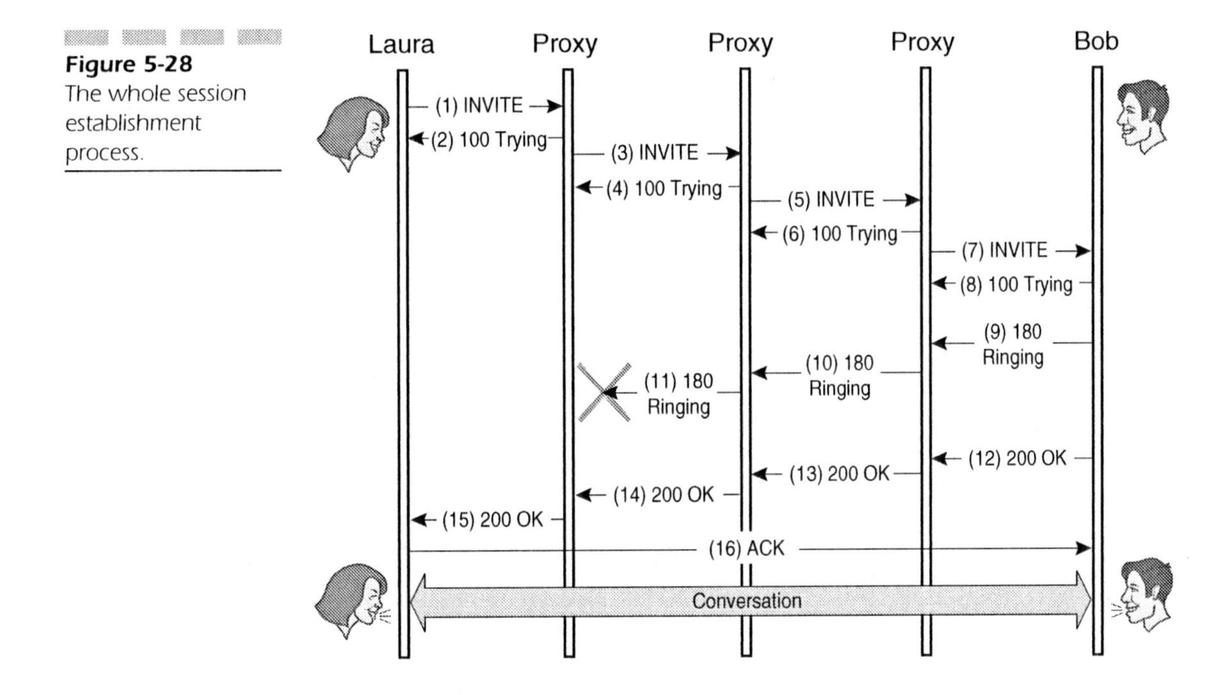

Figure 5-28 The whole session establishment process.

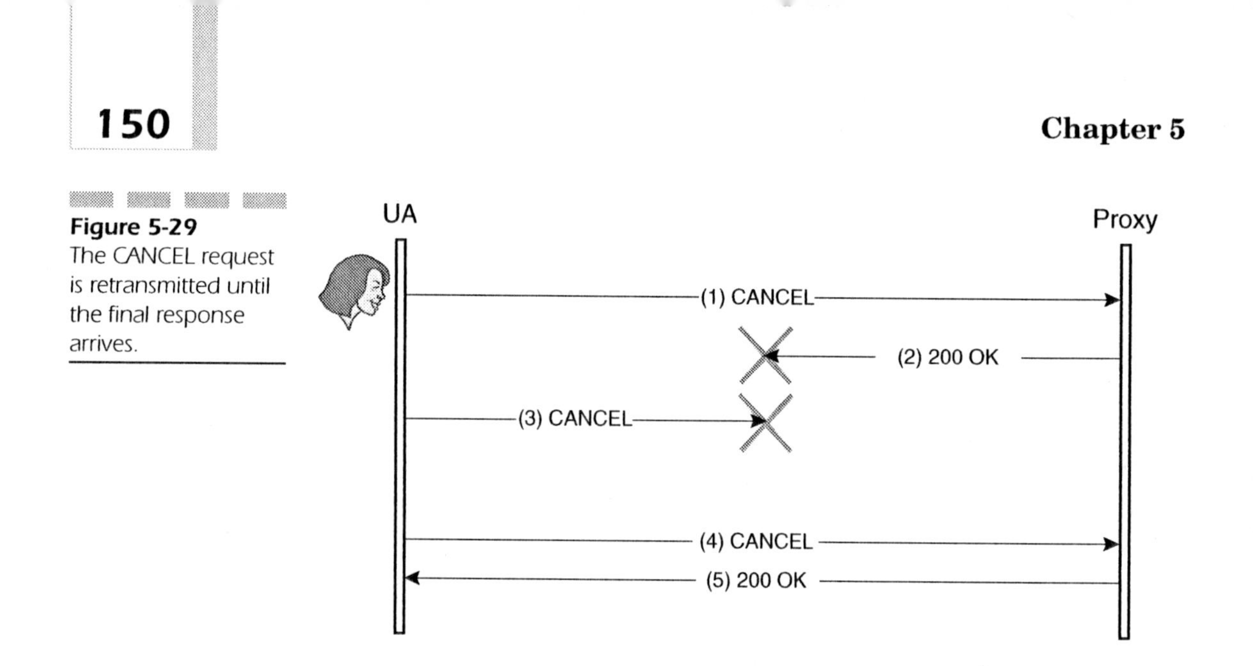

Figure 5-29 The CANCEL request is retransmitted until the final response arrives.

Other Transactions

In the case of INVITE, ACK, and CANCEL requests, SIP provides reliability mechanisms suitable for the characteristics that all three exhibit. The remaining SIP requests follow common rules. OPTIONS, BYE, and REGISTER are treated in the same way with respect to reliability. In the next chapter, we will see that this commonality enables the protocol to be extended with new methods; a proxy will apply the common reliability rules to any unknown method. Therefore, regarding reliability, no difference appears between a BYE and any new method.

The common reliability rules also employ the hop-by-hop mechanism used for INVITEs. A UA makes sure the request is received by the next proxy, and then the next proxy ensures that the following proxy in the path receives it, and so on. When the final response comes from the remote end, the proxy will ensure that it is delivered to the UA that originated the request.

For reliable transport, the UA sends the request to the proxy. When the final response arrives, the proxy will return it to the UA also using reliable transport (Figure 5-30). Any provisional response that arrives before the final response is also sent to the UA over the reliable transport protocol.

For unreliable transport, the UA has to ascertain that the proxy receives the request. When the proxy receives a response from the remote end, it has

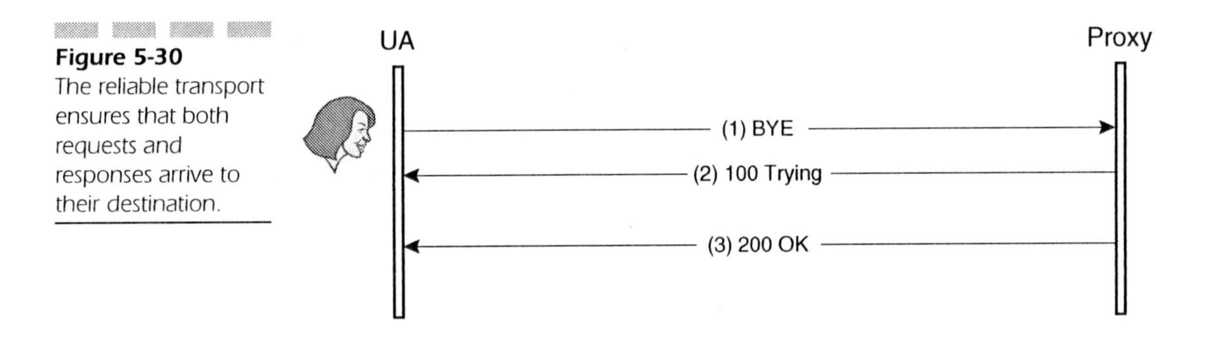

Figure 5-30 The reliable transport ensures that both requests and responses arrive to their destination.

to ensure that the UA also receives it. The UA retransmits the request until the proxy proffers a final response. The proxy retransmits its final response as long as it continues receiving request retransmissions. When retransmissions stop, the proxy determines that the UA has received the final response (Figure 5-31).

Detailed Example

Putting together all these pieces, we can now explore an example of how SIP works in detail. As noted, the examples usually consist of message flows that let us schematize SIP transactions and message exchanges. It is usually not necessary to examine all the parameters of each message to understand a particular scenario. In this scenario, we do so deliberately to display SIP messages with all their headers and parameters.

SIP Call Through a Proxy

In Figure 5-32, Laura calls Bob at his public address, but he isn't there. A proxy server at company.com routes the call to his current location: SIP:Bob@131.160.1.112.

This example comprises three different transactions: INVITE, ACK, and BYE.

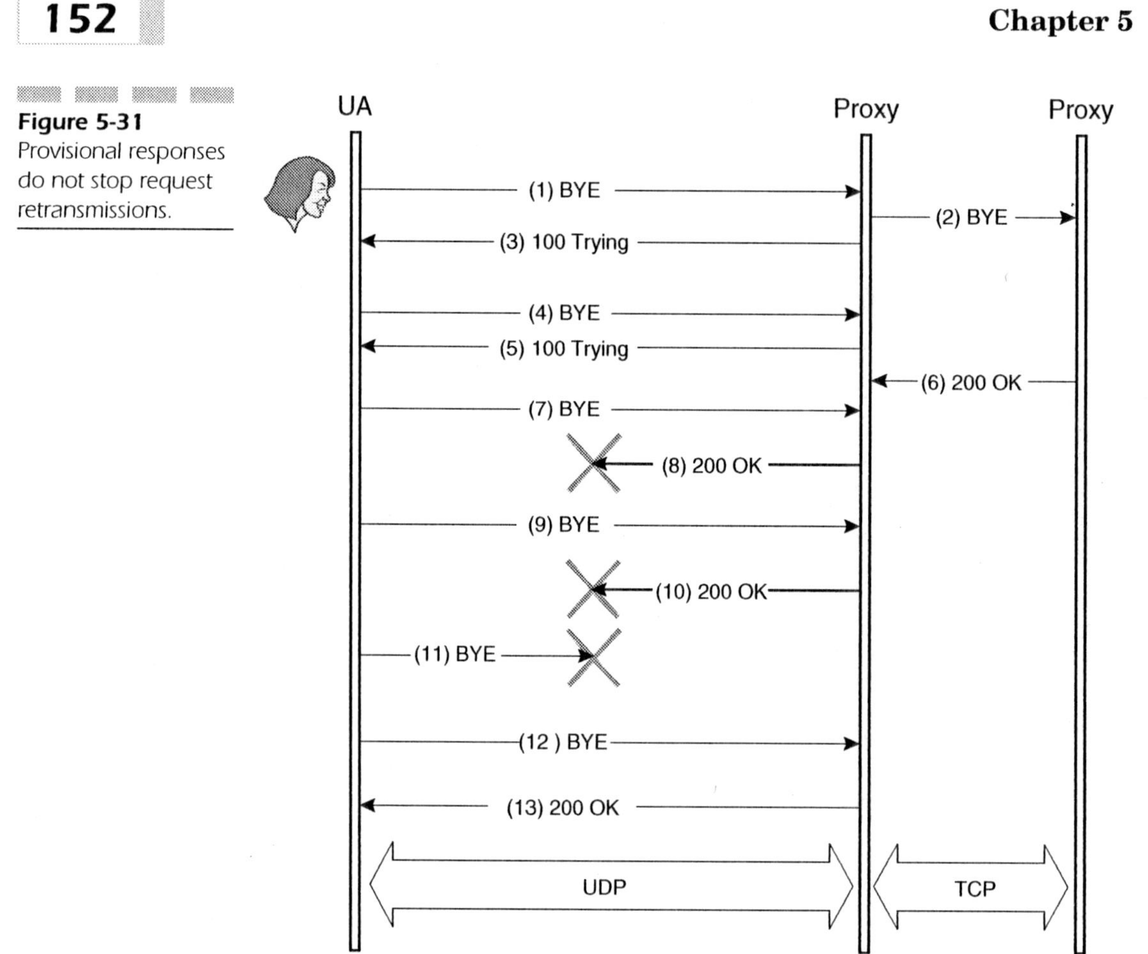

Figure 5-31 Provisional responses do not stop request retransmissions.

INVITE from Laura's UA to SIP Proxy Laura's UA places Bob's public address in the To field and in the Request-URI. It adds a Via header with its address and creates a message body with an SDP session description. Laura wants to receive RTP packets containing PCM voice on UDP port 20002. The request is sent to the proxy at company.com because the domain part of the Request-URI is company.com.

```
INVITE sip:Bob.Johnson@company.com SIP/2.0
Via: SIP/2.0/UDP workstation1000.university.com:5060
From: Laura Brown <sip:Laura.Brown@university.com>
To: Bob Johnson <sip:Bob.Johnson@company.com>
```

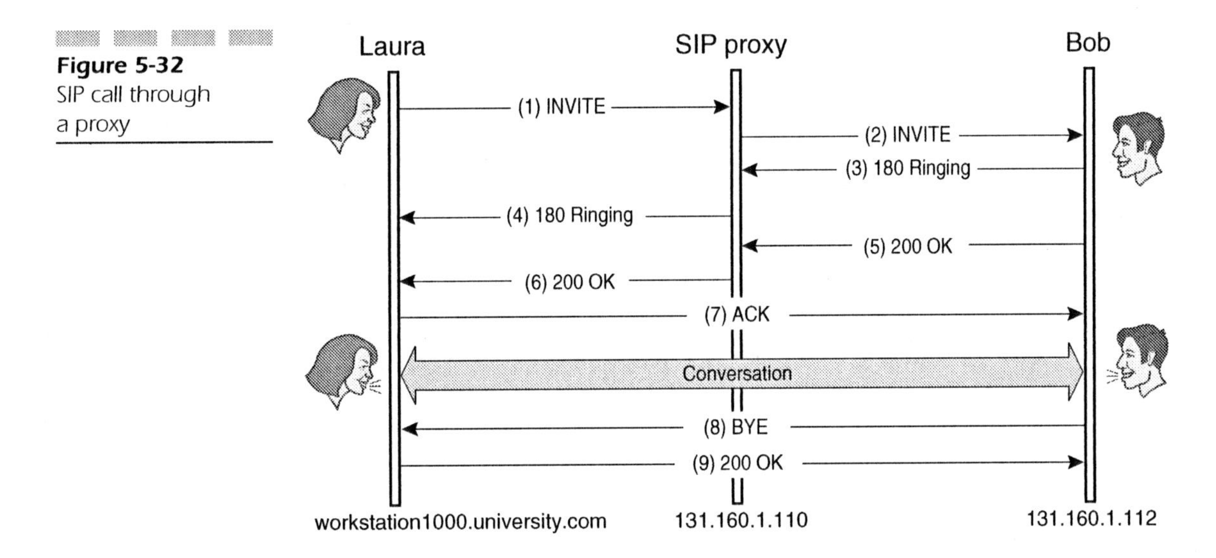

Figure 5-32 SIP call through a proxy

```
Call-ID: 12345678@workstation1000.university.com
CSeq: 1 INVITE
Contact: Laura Brown <sip:Laura@workstation1000.university.com>
Content-Type: application/sdp
Content-Length: 154

v=0
o=Laura 2891234526 2891234526 IN IP4 workstation1000.university.com
s=Let us talk for a while
c=IN IP4 138.85.27.10
t=0 0
m=audio 20002 RTP/AVP 0
```

INVITE from SIP Proxy to Bob The SIP proxy at company.com receives the INVITE request. The host part of the Request-URI reads Bob.Johnson. The proxy knows that Bob.Johnson may be reachable at SIP:Bob@131.160.1.112. Thus, it creates a new INVITE with Bob's location as the Request-URI, adding its address to the request as a Via header: 131.160.1.110. Notice that the message body remains untouched. SIP servers do not typically modify message bodies.

```
INVITE sip:Bob@131.160.1.112 SIP/2.0
Via: SIP/2.0/UDP 131.160.1.110
Via: SIP/2.0/UDP workstation1000.university.com:5060
From: Laura Brown <sip:Laura.Brown@university.com>
To: Bob Johnson <sip:Bob.Johnson@company.com>
Call-ID: 12345678@workstation1000.university.com
CSeq: 1 INVITE
Contact: Laura Brown <sip:Laura@workstation1000.university.com>
Content-Type: application/sdp
Content-Length: 154

v=0
o=Laura 2891234526 2891234526 IN IP4 workstation1000.university.com
s=Let us talk for a while
c=IN IP4 138.85.27.10
t=0 0
m=audio 20002 RTP/AVP 0
```

Provisional Response from Bob to Proxy Upon receiving the INVITE, Bob's UA must initiate alerting, so it returns a provisional response announcing that alerting has begun. The Via headers are copied from the INVITE received. They will ensure that the response traverses the proxy first, 131.160.1.110, and then arrives at Laura's UA, workstation1234.university.com at UDP port number 5060.

Bob's UA adds a Contact header to the response containing the SIP URL where Bob can be reached directly; from now on, subsequent requests will be sent directly from Laura's UA to Bob's UA.

Bob's UA also adds a tag parameter to the To header, naming the SIP UA that Bob is currently using. The tag info helps to differentiate the responses that Laura might get if a forking proxy in the path tried to reach Bob at several locations. To avoid confusing Laura's UA, each of Bob's UAs will have answered with a different tag.

```
SIP/2.0 180 Ringing
Via: SIP/2.0/UDP 131.160.1.110
Via: SIP/2.0/UDP workstation1000.university.com:5060
```

```
From: Laura Brown <sip:Laura.Brown@university.com>
To: Bob Johnson <sip:Bob.Johnson@company.com>;tag=314159
Call-ID: 12345678@workstation1000.university.com
CSeq: 1 INVITE
Contact: Bob Johnson <sip:Bob@131.160.1.112>
```

Provisional Response from Proxy to Laura Upon receipt of this response, the proxy removes the Via header with its own address and sends the response to the address contained in the next Via header. This proxy takes no further action.

```
SIP/2.0 180 Ringing
Via: SIP/2.0/UDP workstation1000.university.com:5060
From: Laura Brown <sip:Laura.Brown@university.com>
To: Bob Johnson <sip:Bob.Johnson@company.com>;tag=314159
Call-ID: 12345678@workstation1000.university.com
CSeq: 1 INVITE
Contact: Bob Johnson <sip:Bob@131.160.1.112>
```

Final Response from Bob to Proxy When Bob accepts the call, his UA returns its SDP session description. It will receive RTP packets on UDP port 41000.

```
SIP/2.0 200 OK
Via: SIP/2.0/UDP 131.160.1.110
Via: SIP/2.0/UDP workstation1000.university.com:5060
From: Laura Brown <sip:Laura.Brown@university.com>
To: Bob Johnson <sip:Bob.Johnson@company.com>;tag=314159
Call-ID: 12345678@workstation1000.university.com
CSeq: 1 INVITE
Contact: Bob Johnson <sip:Bob@131.160.1.112>
Content-Type: application/sdp
Content-Length: 154
```

```
v=0
o=Bob 2891234321 2891234321 IN IP4 131.160.1.112
s=Let us talk for a while
c=IN IP4 131.160.1.112
t=0 0
m=audio 41000 RTP/AVP 0
```

Final Response from Proxy to Laura The proxy server routes the final response in the same way it routed the previous provisional response. In other words, it removes the first Via header and sends the response to the address contained in the next Via.

```
SIP/2.0 200 OK
Via: SIP/2.0/UDP workstation1000.university.com:5060
From: Laura Brown <sip:Laura.Brown@university.com>
To: Bob Johnson <sip:Bob.Johnson@company.com>;tag=314159
Call-ID: 12345678@workstation1000.university.com
CSeq: 1 INVITE
Contact: Bob Johnson <sip:Bob@131.160.1.112>
Content-Type: application/sdp
Content-Length: 154

v=0
o=Bob 2891234321 2891234321 IN IP4 131.160.1.112
s=Let us talk for a while
c=IN IP4 131.160.1.112
t=0 0
m=audio 41000 RTP/AVP 0
```

ACK from Laura to Bob When Laura's UA receives the 200 OK final response, it sends an ACK request. This ACK is sent directly to Bob's UA, whose address is contained in the Contact header just received.

```
ACK sip:Bob@131.160.1.112 SIP/2.0
Via: SIP/2.0/UDP workstation1000.university.com:5060
From: Laura Brown <sip:Laura.Brown@university.com>
To: Bob Johnson <sip:Bob.Johnson@company.com>;tag=314159
```

```
Call-ID: 12345678@workstation1000.university.com
CSeq: 1 ACK
Contact: Laura Brown <sip:Laura@workstation1000.university.com>
```

BYE from Laura to Bob Now Laura is ready to finish the call so her UA sends a BYE request. This BYE request is also sent directly to Bob's UA using the Contact header previously received. Note that the Cseq has been increased. This BYE request belongs to a new transaction.

```
BYE sip:Bob@131.160.1.112 SIP/2.0
Via: SIP/2.0/UDP workstation1000.university.com:5060
From: Laura Brown <sip:Laura.Brown@university.com>
To: Bob Johnson <sip:Bob.Johnson@company.com>;tag=314159
Call-ID: 12345678@workstation1000.university.com
CSeq: 2 BYE
Contact: Laura Brown <sip:Laura@workstation1000.university.com>
```

Final Response from Bob to Laura Bob's UA receives the BYE request, terminates the audio session, and returns a 200 OK response for the BYE.

```
SIP/2.0 200 OK
Via: SIP/2.0/UDP workstation1000.university.com:5060
From: Laura Brown <sip:Laura.Brown@university.com>
To: Bob Johnson <sip:Bob.Johnson@company.com>;tag=314159
Call-ID: 12345678@workstation1000.university.com
CSeq: 2 BYE
Contact: Bob Johnson <sip:Bob@131.160.1.112>
```

The example above shows how all the headers that were explained previously in this chapter work together to establish a voice session. We can already see that SIP protocol opertion is pretty simple and easy to understand.

CHAPTER 6

Extending SIP: The SIP Toolkit

Session Initiation Protocol (SIP) is designed so that its core functionality is present in every implementation. This provides interoperability on a global scale. Every SIP implementation can capitalize on the fact that any other SIP implementation will be able to understand all of the mechanisms described in the SIP *Request for Comments* (RFC) [RFC 2543]. However, some implementations need functionality beyond the core protocol, which means that SIP needs to be enhanced somehow.

SIP is flexible and easy to extend. The community has defined a set of extensions very quickly. Applications with special requirements devise extensions to meet their particular needs. These extensions are implemented in a modular fashion and their use can be negotiated during session establishment, ensuring that a simple *User Agent* (UA), whose sole purpose is to implement the core protocol, will always be able to interoperate with more advanced UAs.

SIP extensions can be seen as the SIP toolkit. Every one of the extensions resolves a concrete problem. In order to solve a big problem, such as how to provide a new service, you can anticipate that it will be necessary to combine the core specification with the suitable extensions.

This chapter, like Chapters 4 and 5, describes existing SIP extensions by first explaining each one's functionality and then outlining how each extension is implemented. Once the reader becomes familiar with the most common SIP extensions, we can devote the next chapter to the architectures and applications that use the SIP toolkit (SIP and its extensions) for service provisioning.

Extension Negotiation

A given SIP application can always assume that another SIP application is able to understand the SIP core protocol. However, it cannot make any assumptions about which extensions the remote end supports. Therefore, a negotiation process is needed in order to determine the extensions that will be used within any given session (Figure 6-1).

During this process, the SIP entities involved negotiate two things: which extensions the remote party supports and which extensions will actually be employed in the upcoming session. For instance, two very advanced SIP UAs might support several extensions, but they will not necessarily use any if the core protocol suffices for the type of session being

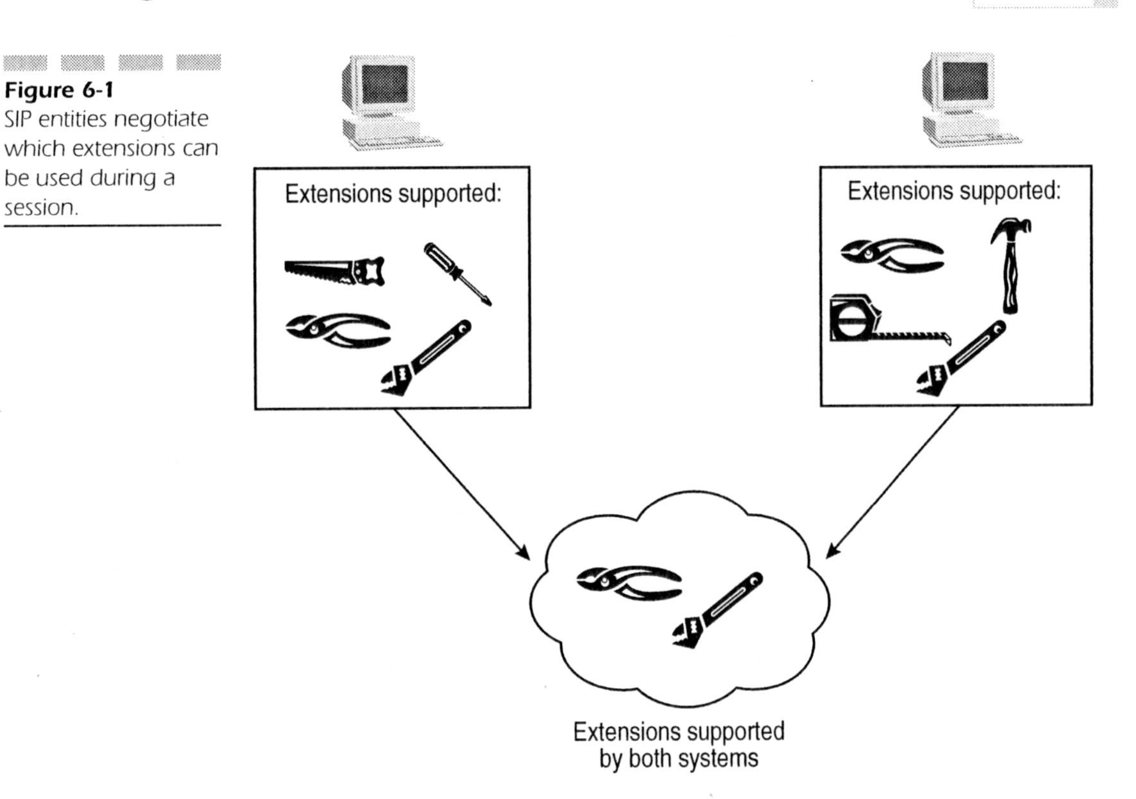

Figure 6-1 SIP entities negotiate which extensions can be used during a session.

established at a particular moment. However, at any other moment, they might decide to use a subset of all the extensions they support. SIP maintains a clear distinction at all times between the extensions that are understood by both parties and those that will be used in a particular session.

How It's Done

The process of negotiating the session's extensions is conducted using two headers: Require and Supported [draft-ietf-sip-serverfeatures]. A client lists all of the extensions it needs to establish a session in the Require header and all of the extensions it supports in the Supported header. The server decides which extensions will be used in the session based on these two headers.

At the other end, the server lists the extensions that will be used in the Require header and the extensions it understands in the Supported header. This way, the server reciprocates the client's information about which extensions it supports.

In Figure 6-2, Bob's UA supports extensions foo1, foo2, foo3, and foo4. Bob wants to use extension foo1 in the session, so he adds it to the Require header of his INVITE. Laura wants to use extension foo2 as well. She knows that Bob's UA supports it because it appears in the Supported header of the INVITE. Therefore, she adds foo2 to the Require header. Besides that, she tells Bob's UA that her UA also supports foo4 and foo5. Now the Require header in the 200 OK from Laura contains the extensions that will actually be used through the session. The information contained in the Supported header might prove useful later if Bob decides mid-session to use the extension foo4 as well; he is forearmed with the knowledge that it's available and can make a user's decision to expand his session capabilities. The extensions foo3 and foo5 will not be used in the session because they are only understood by one of the end systems.

Design Principles for SIP Extensions

Effective design for new SIP extensions must follow certain rules. Some design principles for SIP extensions have been defined [draft-ietf-sip-guidelines] to ensure that new extensions do not change the spirit of SIP.

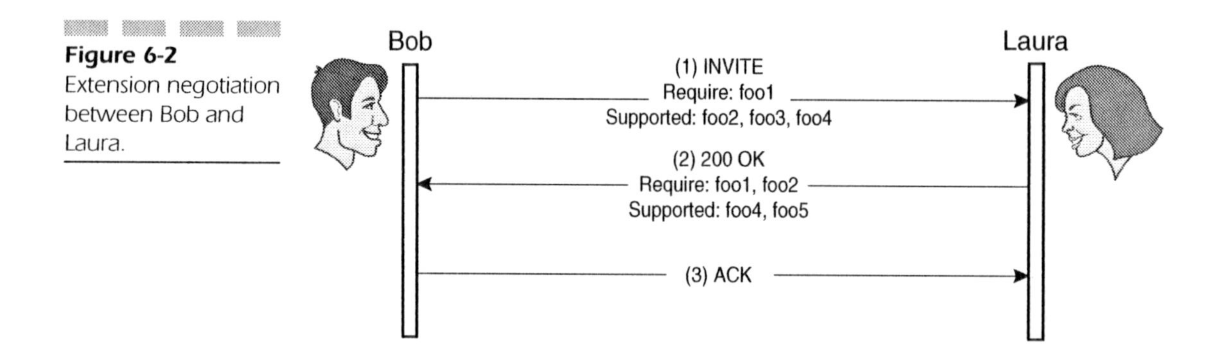

Figure 6-2 Extension negotiation between Bob and Laura.

New proposed extensions in the form of Internet drafts are carefully analyzed in the SIP working group before they are accepted as standard SIP extensions. For those readers who will be designing extensions, let's enumerate the characteristics that an extension will have to exhibit in order to be accepted by the SIP community.

Do Not Break the Toolkit Approach

One of the great advantages of SIP is that it fits into the multimedia conferencing architecture of the Internet. SIP is part of the *Internet Engineering Task Force* (IETF) toolkit for multimedia. It does what it was designed to do and makes use of other protocols for other tasks (such as *Session Description Protocol* (SDP) for session description). Extensions to SIP should not broaden the scope of SIP so that SIP is used for tasks that are handled better by other Internet protocols, even if, at the moment, it appears SIP would do the job.

For example, one might want to use SIP to download Web pages from a server. HTTP already exists for that purpose. Therefore, it would not be wise to create SIP extensions to try to cover HTTP functionality. SIP should be used to locate a particular SIP entity and deliver an object (such as a session description), possibly followed by a negotiation. SIP should be used for applications that leverage the user mobility, object delivery, and negotiation provided by SIP. All other applications fall outside of the scope of SIP. If we tried to use SIP to resolve every problem it might solve, the protocol would quickly become huge and complex, which is the opposite of the IETF philosophy that created it in all its elegance. The IETF standardization process ensures that SIP is kept simple and manageable.

Peer-to-Peer Relationship

SIP entities usually have a peer-to-peer relationship. When a server receives a request from a client, it performs some tasks and then returns a response with the result of the request. The client does not continuously send orders to the server telling it how to proceed. Thus, SIP is not really effective in a master/slave architecture where the master has a lot of control over the slave.

SIP extensions should not be used to provide such control functionality, which is already provided by more suitable protocols such as H.248

[RFC3015]. The peer-to-peer relationship between SIP entities, conversely, makes the protocol highly suitable for inter-domain communications. Master/slave protocols have been proven ineffective for inter-domain communications, where domain owners typically want to prevent the owner of a different domain from controlling their resources.

Independence from Session Type

SIP separates session establishment from session description. This separation has to be maintained whenever extensions are added to the core protocol in order to ensure that extensions will be future-proof (that is, in SIP's case, capable of handling any type of session). For example, a new extension should define how SIP interacts with generic *Quality of Service* (QoS) mechanisms, but should not define how SIP can be used in conjunction with SDP and *ReSerVation Protocol* (RSVP) in order to provide QoS. That's not to say that the latter wouldn't be useful in some cases, but in every case, the generic mechanism would be defined as a SIP extension and the concrete application of the generic mechanism to SDP and RSVP would be described in an informational document only.

It's important to keep this in mind because although SIP can be used to establish all types of sessions, SIP development to date has been quite focused on *Voice over IP* (VoIP) applications. This kind of focus is not unusual in developing protocols, but we have to guard against the natural tendency to design extensions that will be applicable only in a VoIP context. Presently, SIP extensions are general enough to cover different types of sessions, even if their current use is a VoIP service.

Do Not Change Method Semantics

The purpose of a SIP request is defined by its method (for example, a BYE request is intended to terminate a session). Thus, it is possible to know the purpose of a request at first glance by inspecting its method. Headers and parameters give more information about the request, but the general purpose of the request is not changed by the contents of any header.

This design rule remains when new SIP extensions are planned. Let us look at an example of an extension that would not be accepted by the IETF. The following extension breaks the rule of identifying the purpose of a request by its method and therefore would be a bad idea.

One might define a header called Real-Purpose to be carried in INVITE requests as follows:

```
Real-Purpose: Tell me your capabilities
```

An INVITE with this header would be used to query a remote system about its capabilities rather than to establish a session.

However, the purpose of INVITE is not to query about capabilities. If a system wants to query about capabilities, it should use the OPTIONS method, which is defined specifically for that purpose.

Therefore, whenever a request with a new functionality is needed, the SIP community creates a new method. They do not try to change the semantics of an existing one.

Extensions to SIP

We have already seen which common characteristics can be expected from every SIP extension. In this section, we will analyze some SIP extensions that have been proposed. We will describe the functionality provided by the extension and outline how it is implemented, and will try to keep these two concepts as separate as possible so the reader can readily distinguish them.

The SIP Toolkit

The core protocol taken together with its extensions can be seen as a toolkit for creating services. In order to design a new application or create a new service, the designer picks the extensions needed and combines them. Thus, extensions are sufficiently general mechanisms that can be used in a variety of different services. We'll follow the same bottom-up approach to describe SIP extensions, postponing a discussion of specific applications and architecture that have been implemented using different extensions until the next chapter.

Reliable Delivery of Provisional Responses

Core SIP ensures that the initiator of a session is informed when the remote party agrees to join the session. Short of actual acceptance, though, reports on the progress of session establishment are not considered valuable.

However, some applications need this type of information. For example, suppose the support department of a company implements a queue for incoming calls. When all of the clerks are busy, calls from customers are placed in the call queue. As the next clerk becomes available, he or she answers the first call in the queue.

Customers calling this service will insist on being informed about their progress in the queue. They can be informed either by displaying a message on the screen of the caller's SIP phone (such as "You are now the second in the queue") or by playing a voice message for the caller. The relevant extension for providing this service is the reliable delivery of provisional responses.

How It's Done Information about how session establishment is progressing is carried in provisional responses for an INVITE request. The provisional response 180 Ringing from the callee to the caller indicates that the callee is being alerted. The provisional response 182 Queued indicates that the call has been placed in a queue.

SIP does not transmit provisional responses reliably. When a *User Agent Server* (UAS) returns a provisional response to the *User Agent Client* (UAC), it traverses the same proxies as the INVITE did but in the opposite direction. Even if proxies in the path forward this provisional response to the UAC, because UASs do not retransmit provisional responses, any router in the network may discard the IP datagram containing the SIP response. So, it might come about that the UAC never receives the provisional response (Figure 6-3).

[Draft-ietf-sip-100rel] is the SIP extension defined to provide the reliable transfer of provisional responses. UASs transmitting reliable provisional responses retransmit until a message from the UAC is received acknowledging reception. This mechanism is similar to the one used by core SIP for 200 OK responses, which are retransmitted by the UAS until the ACK is received.

A new method was defined to acknowledge the reception of provisional responses. This new method is called *Provisional Response ACK* (PRACK). A UAS stops retransmitting a provisional response upon reception of a PRACK from the UAC.

The PRACK request belongs to a different transaction than the INVITE request. Thus, the UAS also must send a response to the PRACK. In Figure 6-4, the UAS sends a 200 OK response to the PRACK, indicating that the PRACK request succeeded.

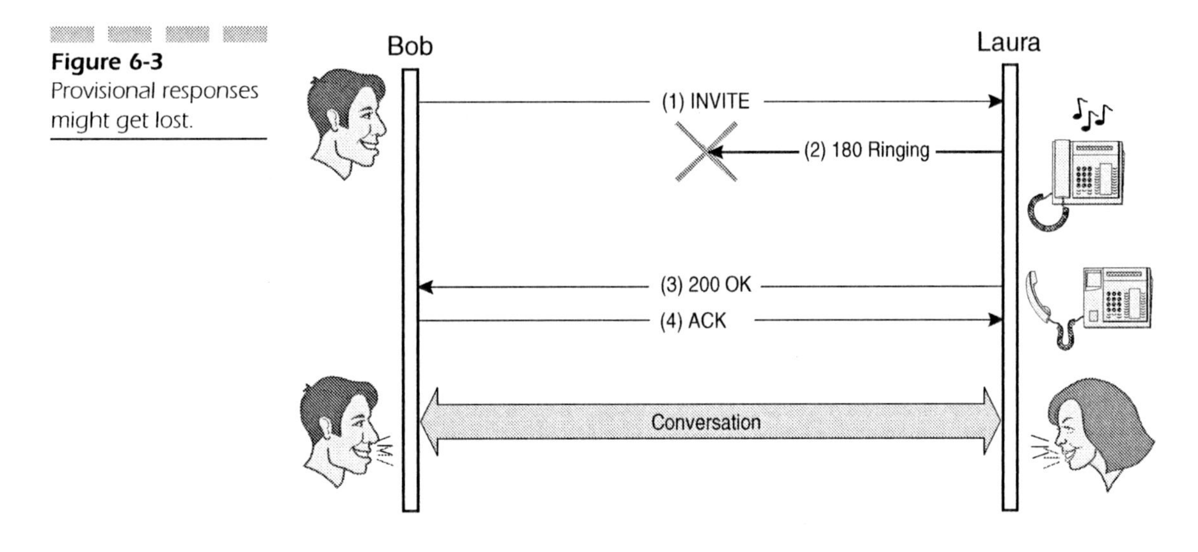

Figure 6-3
Provisional responses might get lost.

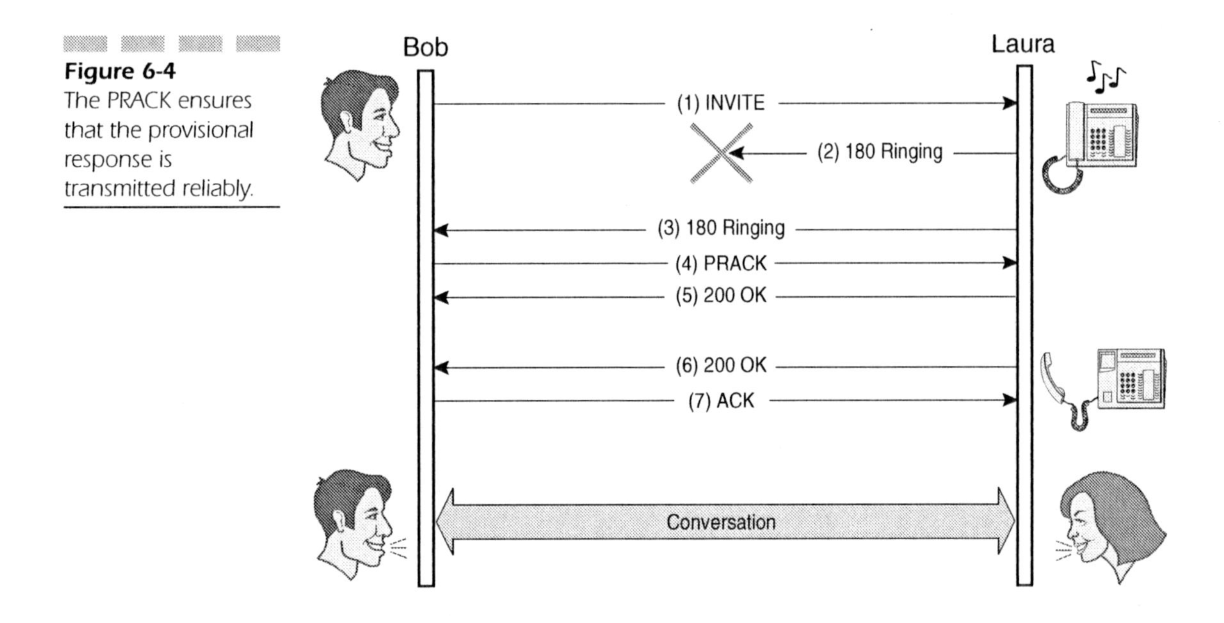

Figure 6-4
The PRACK ensures that the provisional response is transmitted reliably.

Figure 6-5 contains the example that was previously described. A customer is informed about his position in the call queue through reliable provisional responses. His SIP UA displays his new position on the screen every time a new response arrives.

PRACK is an example of the SIP working group that chooses a general solution over a more efficient but less general one. This decision is in line with the IETF paradigm and the toolkit approach. The 200 OK response for the PRACK is really useless from a reliability point of view. The UAC knows that the PRACK has arrived at the UAS because the retransmissions of the provisional response stopped. So why is the 200 OK for the PRACK sent?

The answer to this question is one of the key features of the Internet: end-to-end protocols perform end-to-end functions better. SIP proxies understand the basic SIP methods. We saw in Chapter 5 that a proxy receiving a method that it does not understand routes it by following the same rules as for BYE requests. The proxy basically forwards the request and if it is stateful, waits for a final answer.

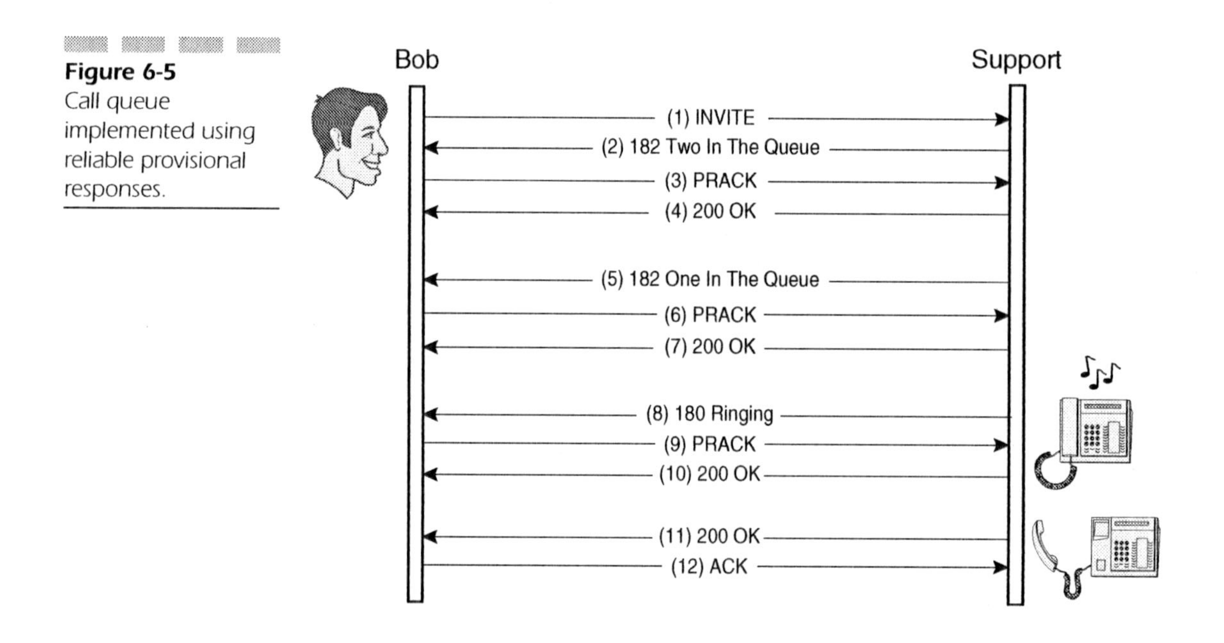

Figure 6-5 Call queue implemented using reliable provisional responses.

Making PRACK into a request-response transaction enables proxy servers that are not PRACK aware to route this new method properly. If PRACKs had no response, proxies would have to learn a new method in order to apply special routing that doesn't wait for a response.

This example shows how a new service can be implemented without modifying anything in the network. Two terminals that implement the PRACK method are able to exchange reliable provisional responses over the same network of SIP servers that was implemented when PRACKs did not exist. Those SIP servers will not understand PRACK because it was developed when they were already installed in the network, but they can still route PRACKs properly. Therefore, end systems can implement new services and use them over an existing SIP network without requiring an upgrade. This is one of the reasons why the pace at which services are created for SIP is much faster than for networks other than the Internet.

Mid-session Transactions That Do Not Change the State of the Session

We have already seen several examples of how SIP is used to establish sessions. Once a session is established, the core SIP specification provides a means to change the parameters of the sessions through re-INVITEs and a means to terminate the session through a BYE. Certainly, in some situations, the parties involved in a session may need to exchange information that does not influence the state of the session in any way. However, core SIP does not provide a means of sending information to the remote party without changing session state. In this case, the means proves to be an extension for mid-session information. Exchange of this kind of information can be typically found when SIP interworks with other signalling protocols. One such case is billing information, where the session provider regularly needs to transfer information related to the session that does not modify its parameters. SIP can carry it only by means of extension.

How It's Done A new SIP method called INFO was defined [RFC 2976] to provide this functionality. INFO transports mid-session information that does not affect state. INFOs typically carry information in the message body that is transferred from end to end between UAs. Figure 6-6 illustrates Bob and Laura exchanging INFOs during an ongoing session.

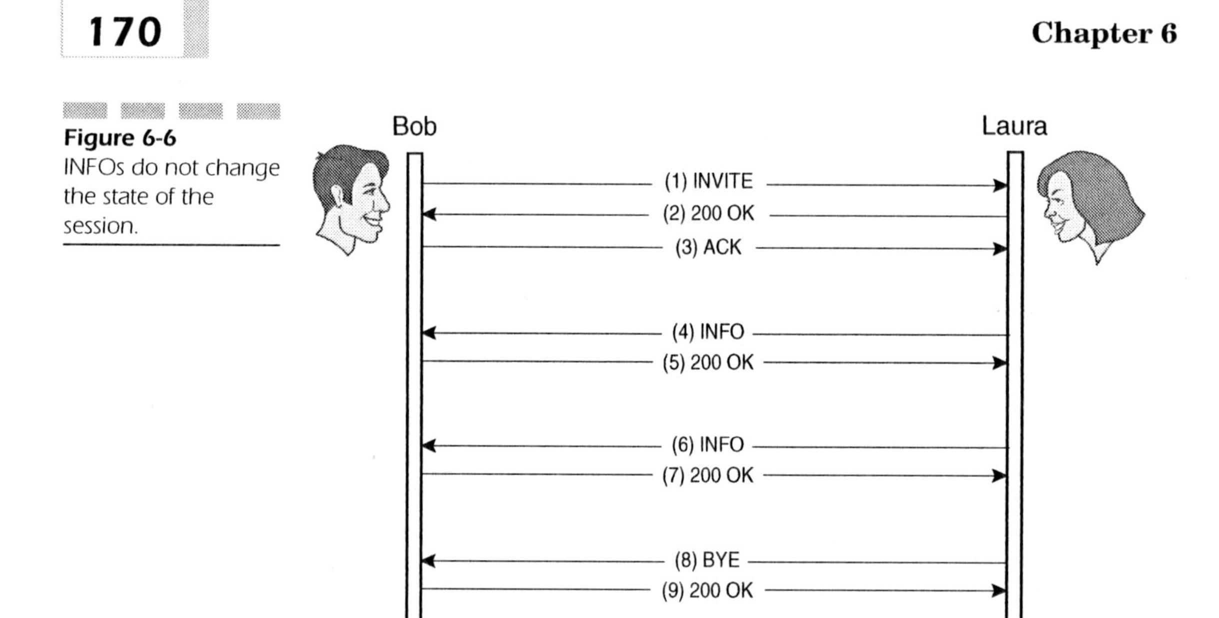

Figure 6-6 INFOs do not change the state of the session.

Multiple Message Bodies

You already know that SIP can locate a user and deliver a session description to the user at his or her current location. However, sometimes it's important to deliver more than a session description. For instance, Bob has a SIP UA in his laptop. When Bob receives a call, apart from receiving the session description that will enable him to establish the session, he finds it useful to see who is calling. SIP can deliver, at session initiation, the caller's photo along with the session description, which will be displayed on Bob's laptop. Bob now has the option of accepting or refusing the call based on the identity of the caller.

How It's Done In order to be able to display the photo of the caller, Bob's UA needs to receive a file with the photo (such as Laura.jpg) or a URL where Bob's UA can retrieve the photo (such as http://www.university.com/~laura/photos/laura.jpg). If a URL is received, Bob's UA uses his Web browser to display the photo. In order to receive this information, whether it turns out to be a file or a URL, Bob's UA needs to receive an INVITE carrying a message body with two parts: an SDP session description and a photo (Figure 6-7).

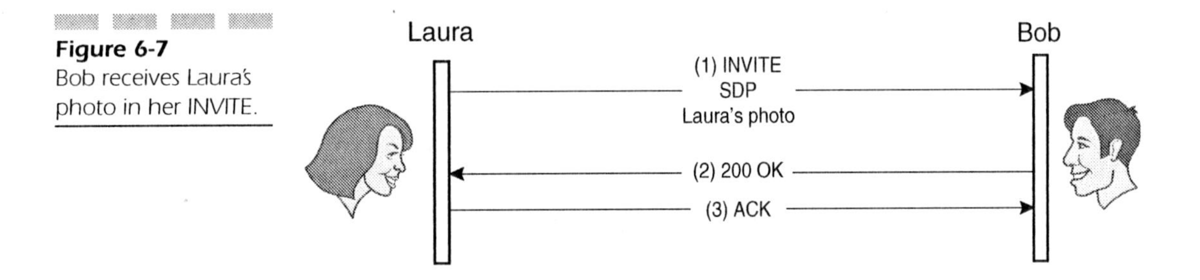

Figure 6-7 Bob receives Laura's photo in her INVITE.

Core SIP accepts multipart message bodies in *Multipurpose Internet Mail Extensions* (MIME) [RFC 2045] format. SIP does not need to be extended in order to support them. Again, SIP uses the same format to send multipart bodies that e-mail systems use to send attachments. This shows the toolkit approach at work so that new services can be built on existing mechanisms.

Instant Messages

To establish a session, SIP delivers a session description to the caller. If instead of a session description, SIP is used to deliver a human-readable message, an instant-messaging system can be trivially implemented with SIP. The main difference between instant messages and e-mail is that instant messages are short and are delivered immediately to the user. E-mails are usually longer and they are stored in the user's inbox until he or she accesses it to read them. It is also common to maintain interactive text conversations using instant messages, but much less so with e-mail where responses aren't always sent within a short period of time.

Many applications use instant messages. For example, several network games let players send short messages to their opponents while play is in progress. Instant messages can also be used in combination with voice. Say Bob is on the phone with Laura trying to explain the correct spelling of a word to her. A short message is much faster than speaking all the letters of the word one by one. SIP with some extensions can provide simple instant-messaging services.

How It's Done A new method called MESSAGE was defined [draft-rosenberg-impp-im] to carry in its body the message that the sender wrote. The big advantage of using SIP for delivering instant messages is that proxies route a MESSAGE request as a BYE request. Therefore, the same infrastructure deployed for establishing multimedia sessions can be reused for providing instant-messaging services without any modifications. Proxies do not need to be aware of this new service in order to route MESSAGE requests.

Figure 6-8 shows Bob and Laura in the midst of a conversation. They establish a SIP session using an INVITE and begin talking. While they're discussing vacation plans, Bob needs to explain to Laura how to spell a word. He sends her an instant message and then they resume conversing by voice.

Automatic Configuration of UAs

One of the main features of SIP is that it provides user mobility through registrations. UAs register their current location to SIP servers so that these servers can route INVITE requests correctly. It is essential for a SIP UA to know which server is handling a particular domain in order to register with that SIP server.

The more mobile the user, the more difficult is for him or her to remember all of the data needed to configure his or her SIP UAs to work in each different domain where the user can be reached.

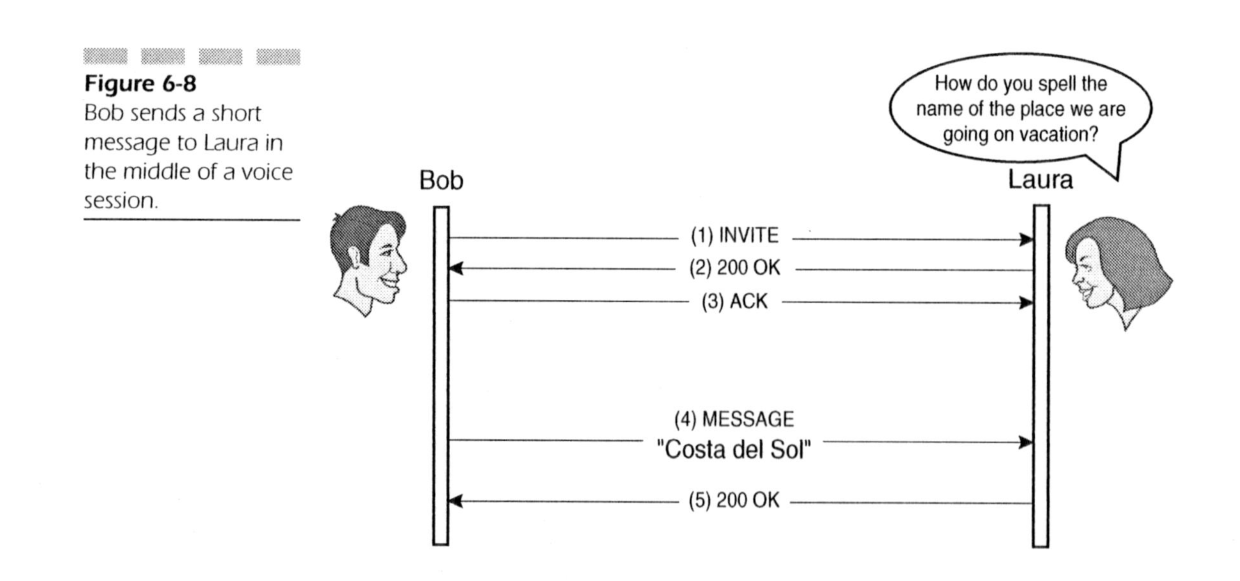

Figure 6-8 Bob sends a short message to Laura in the middle of a voice session.

Bob, for example, has a SIP UA in his laptop for making VoIP calls. When he is working in his office, he tends to use the SIP server at company.com. He knows the name of the domain, company.com, so he configures his SIP UA in order to send REGISTERs to that SIP server.

Bob usually works in the university in the afternoon. When he gets there, he'll have to change the configuration of his SIP UA. He sends the REGISTERs to the proxy at university.com, and also configures this proxy as the outbound proxy. All requests from his UA will be sent first to this proxy because the university has a firewall that will only admit SIP messages have traversed the proxy at university.com. This security proxy works much like the HTTP proxy used by many companies to access the Internet.

Bob is accustomed to working in both the university.com and company.com domains, so he knows the configuration data by heart. However, in any other domain, he probably has no idea about how to configure his UA.

This week Bob is visiting another university where he is collaborating with a colleague. When Bob connects his laptop to the Internet, he does not know his current domain name or his outbound proxy. He will most likely have to hunt down the network administrator of the university and ask for this information in person.

Mechanisms are available for SIP server discovery that can make Bob's life easier when he is traveling around trying to use his SIP UA. With these mechanisms, Bob would not need to bother configuring his SIP UA. Just as an IP address is automatically assigned to his laptop when he connects to the Internet, the parameters needed to configure his SIP UA can be automatically delivered to his laptop.

How It's Done This configuration process can be made automatic through a couple of ways. They are actually extensions to protocols other than SIP, but we consider them important enough for SIP applications to mention them.

One option is to use *Dynamic Host Configuration Protocol* (DHCP) [RFC 2131] to retrieve the domain name of the SIP server. In the same way a laptop uses DHCP to retrieve its IP address from a DHCP server, a SIP UA can also retrieve the domain name of a SIP server using DHCP [draft-ietf-sip-dhcp]. Figure 6-9 shows how Bob retrieves both his IP address and his SIP domain name from a DHCP server.

A more advanced UA can use the *Service Location Protocol* (SLP) [RFC 2608] to find a SIP server with certain characteristics. SLP provides server location that matches up server capabilities with user needs [draft-kempf-sip-findsrv].

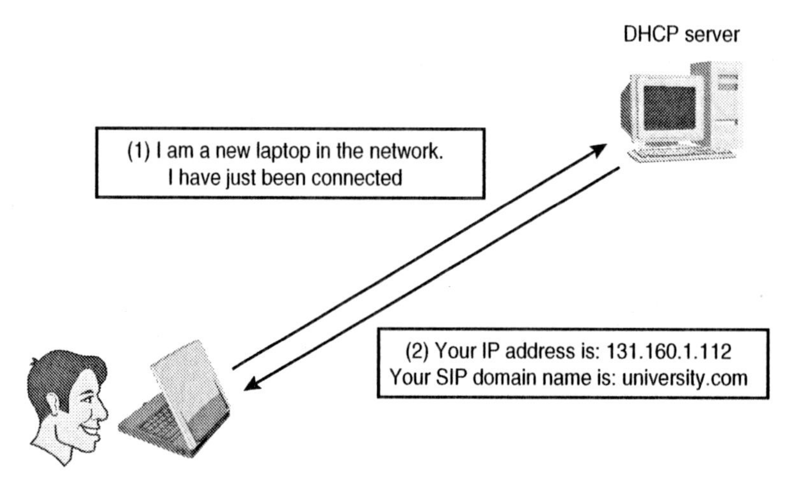

Figure 6-9 Bob's laptop obtains an IP address and a SIP domain name from the DHCP server.

Preconditions to Be Fulfilled Before Alerting

In all the examples we have seen so far, UAS has alerted the user upon reception of an INVITE. If the user accepted the invitation, the session was established. This model works just fine for sessions that can be established quickly. In an audio session, for instance, once the user accepts the call, he or she can hear the caller's voice almost immediately. The model also works for sessions whose requirements for establishing time are not especially tough. For instance, if a user receives an invitation (INVITE) to join a gaming session, he or she does not expect to be able to begin playing as soon as he or she presses the accept button. The user expects (and counts on) some time to elapse between accepting and to joining the game. The application, among other tasks, will probably have to load some game scenarios into memory and synchronize positions on each player's map of different players before the game can start.

The model is unsuitable for sessions that have tougher requirements, however. A good example is a telephone call placed over the *public-switched telephone network* (PSTN). When I receive a call on my PSTN phone, I expect to speak as soon as I pick up. It would be annoying and probably unacceptable if I had to wait so much as 5 seconds between answering the phone and talking.

In the Internet, establishment time increases dramatically if the session requires a certain QoS and/or a certain level of security. Establishing a secure channel between two end users and providing QoS (using RSVP for instance) takes time.

But even if an extremely fast resource reservation mechanism that could fulfill a variety of applications requirements was available, it would still be impossible to know beforehand whether the network will grant the necessary session QoS or not. If the session is established and the network does not grant the necessary QoS, the session fails. In an audio session, this would mean that a user answers a SIP phone that is ringing and finds no call established—a case of ghost ringing.

This problem can be resolved either by changing the user's expectations or meeting the requirements. In the early days of cellular telephony, it was the former. Users of the fixed telephony network were accustomed to very short delays between the moment they finished dialing a telephone number and the first rings on the caller's phone. Cellular users had to wait significantly longer for their calls to be established. However, they were willing to accept this delay in exchange for a new service feature: mobility. A typical user of SIP devices might accept longer establishment delays because he or she can trade off a little patience for a wide range of new services.

As we all know, sometimes it is necessary to simply fulfill the user's requirements. Users who only want voice sessions are most likely to refuse to use a new technology (SIP) if it performs worse than the one they already use.

For these occasions, the SIP toolkit includes a way to meet preconditions for a session, such as security and QoS, before alerting the user. This mechanism ensures that everything is ready for establishing the session when the user agrees to participate in it.

How It's Done A new method called *preCOnditions MET* (COMET) was defined [draft-ietf-sip-manyfolks-resource]. COMETs are sent to indicate that all preconditions are met and session establishment can proceed. (See Figure 6-10 for an example of how to use this method.)

Bob wants to establish a session with Laura that requires QoS. The INVITE he sends her will contain QoS preconditions. Bob specifies that he does not want Laura to be alerted until the network has granted a certain QoS. If this QoS can't be provided, Bob would rather let the session lapse. So Laura's UA receives the INVITE and sends back a "183 Session

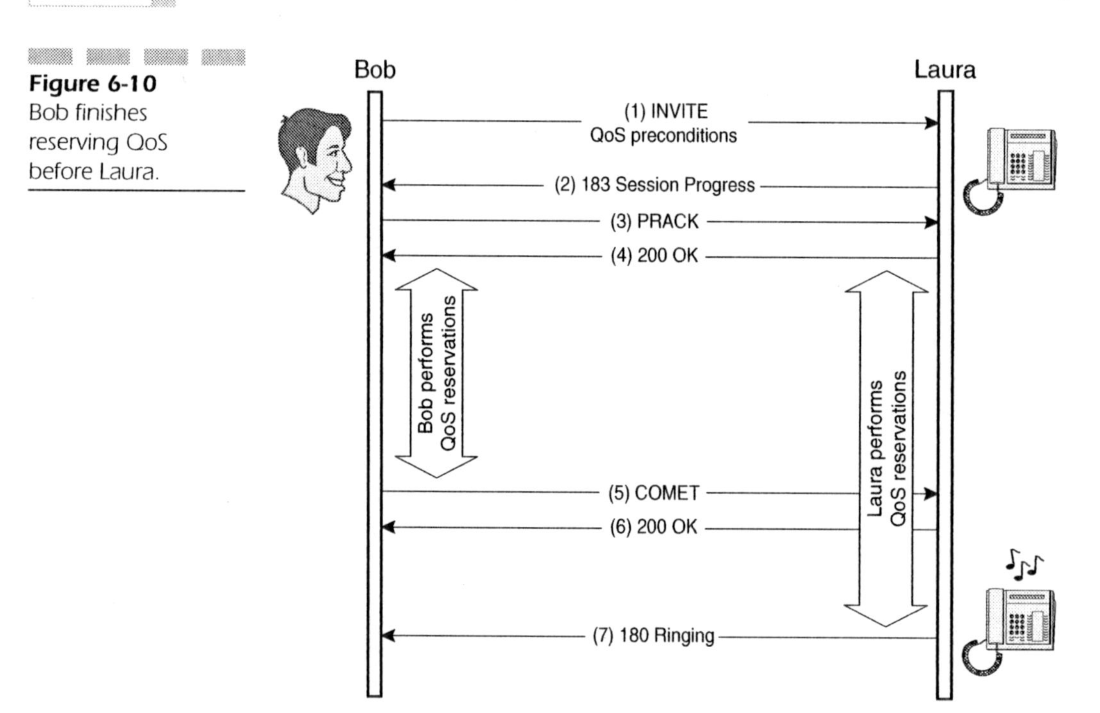

Figure 6-10 Bob finishes reserving QoS before Laura.

Progress" response with Laura's QoS preconditions for the session (this response is transmitted reliably using the techniques described previously in this chapter). Then Bob and Laura perform resource reservation (using RSVP for instance). Bob waits until he finishes reserving resources to send a COMET to Laura. When Laura's UA has also completed resource reservation, it resumes the session establishment process by alerting Laura.

Figure 6-11 illustrates what happens if Laura's UA finishes resource reservation ahead of Bob's. Laura's UA waits until a COMET arrives from Bob. Once the COMET is received, it alerts Laura.

Caller Preferences

We have seen SIP servers handle requests in many different ways. A SIP server, for instance, can choose between performing parallel or sequential searches, and between forking or trying a single location. The core SIP specification leaves these decisions up to the administrator who configures the server.

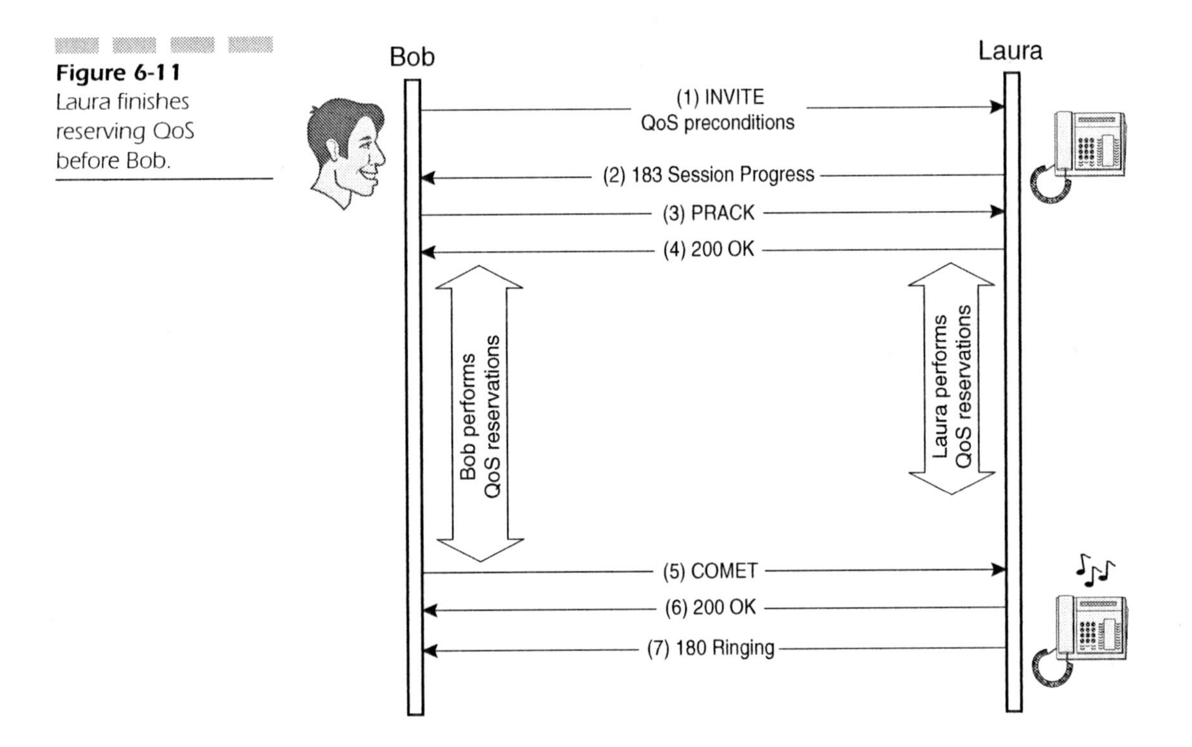

Figure 6-11 Laura finishes reserving QoS before Bob.

The caller, however, might have different preferences. I might want all my requests to be forked in parallel rather than sequentially to save time, whereas someone else might want his or her requests to be forked sequentially to save sanity. Most people will want their fixed phone rung first and, if nobody answers, their mobile phone rung next. The concept of all my contact devices being pinged at once is an attractive one, but the concept of being pinged by all my contact devices at once is distinctly unpleasant.

Besides influencing how the caller's requests are handled, a certain caller might also be interested in defining which kind of terminal he or she wants to reach. For instance, when Bob is calling Laura to talk about the opera they saw last Saturday, he does not want to reach Laura at work. He prefers to reach Laura at her private line. On the other hand, when Bob's boss calls him to talk about work-related matters, he does not want to reach Bob at his personal SIP phone. When Bob wants to speak with Laura for a long while, he does not want to reach her at her mobile SIP phone because calling this phone is more expensive than calling her fixed SIP phone. Thus, Bob wants his INVITEs to reach Laura's fixed terminal rather than her mobile one.

Knowing caller preferences is also useful for providing services. A user calling the tax office might speak English rather haltingly. Therefore, he wants his INVITE to reach a Spanish-speaking representative to help him with his tax declaration.

Some extensions to SIP enable callers to describe how their requests should be handled and which kind of SIP UA they would like to reach. However, note that the preferences of the caller have to interact with the configuration of the server. Thus, if a server is configured to divert incoming calls to a mobile and the caller states that he or she wants the INVITE to reach a fixed terminal, a mismatch must be resolved. In this situation, the caller would have to clarify in the request whether his or her preference for fixed terminals can be overruled or whether he or she would not accept connection to a mobile terminal at all.

How It's Done In order to achieve this functionality, three new headers and new Contact header parameters were defined [draft-ietf-sip-caller-prefs]. The new headers are Accept-Contact, Reject-Contact, and Request-Disposition. New parameters for Contact headers describe the user's terminal.

Figure 6-12 demonstrates the usage of these new headers and parameters. Bob sends a REGISTER to his server stating that he is available at

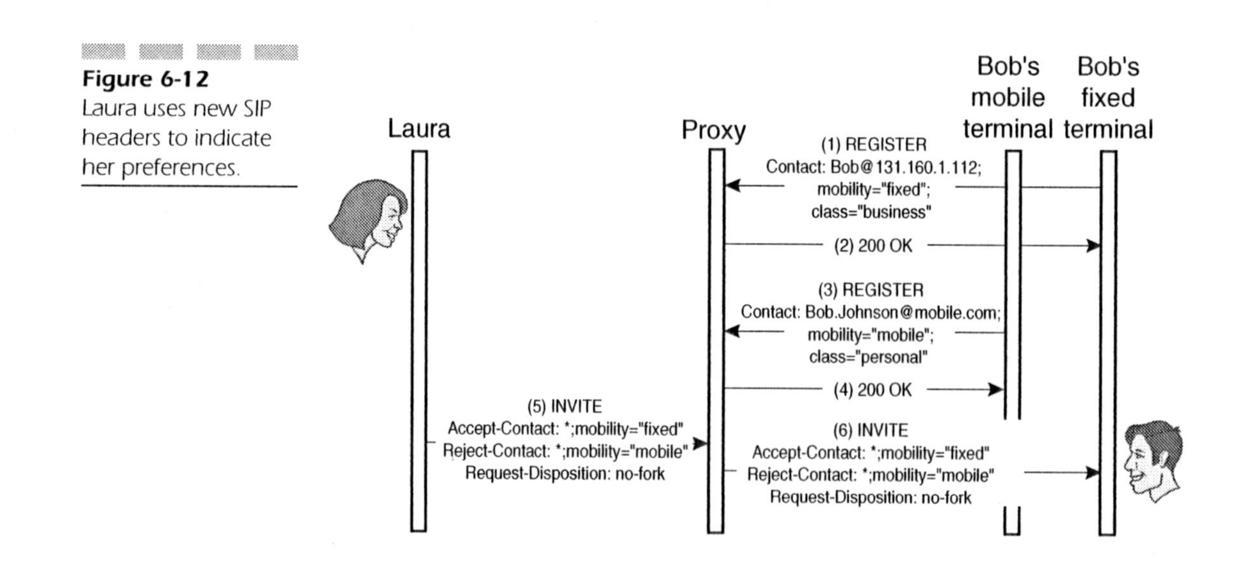

Figure 6-12 Laura uses new SIP headers to indicate her preferences.

SIP:Bob@131.160.1.112 and specifying that this is a fixed terminal used for business. Therefore, the Contact header of his REGISTER looks like the following:

```
Contact: sip:Bob@131.160.1.112; mobility="fixed"; class= "business"
```

Bob sends a second REGISTER saying that he is also available at SIP:Bob.Johnson@mobile.com. He specifies that this is a mobile terminal used for personal matters. The Contact header of this REGISTER looks like the following:

```
Contact: sip:Bob.Johnson@mobile.com; mobility="mobile"; class=
"personal"
```

Now Laura wants to call Bob. She sends an INVITE containing her preferences. Today she has to reach Bob on a fixed terminal and does not want under any circumstances to reach Bob on his mobile. To express these preferences, Laura uses the Accept-Contact and Reject-Contact headers.

```
Accept-Contact: * ; mobility="fixed

Reject-Contact: * ; mobility="mobile"
```

The * is a wild card. These headers together mean that Laura agrees to reach any fixed address while she refuses any mobile address. Laura also adds a Request-Disposition header to her INVITE to prevent her INVITE from forking:

```
Request-Disposition: no-fork
```

Asynchronous Notification of Events

In the examples discussed so far, SIP signalling has been triggered by an action performed directly by a person wishing to establish, modify, or terminate a particular session. Bob triggers SIP signalling (an INVITE) when he picks up his SIP phone to call Laura and also when he hangs up (a BYE). However, other events besides establishing, modifying, or terminating a session can trigger SIP signalling. The SIP *event notification framework* enables SIP to inform users about a variety of events in which they've previously indicated an interest via signalling. This event notification mechanism is a powerful tool for building services. Say Bob calls Laura and she is busy. Even if he's in a hurry, he does not want to keep trying her number every minute to determine when she becomes available. Instead, he wants news of her availability to be delivered to him.

Another example where an event notification mechanism is useful is the implementation of a presence service. A presence service usually consists of a list of your friends and colleagues and their current availability to communicate with you. Bob can see in his buddy list whether Laura is willing to receive telephone calls at a certain moment or whether she just wants instant messages.

SIP can also be used to build this presence service. Every time Laura's status changes, Bob is notified. This way Bob's buddy list will reflect changes in Laura's status.

How It's Done Two new methods were defined to provide asynchronous event notification: SUBSCRIBE and NOTIFY [draft-ietf-sip-events]. SUBSCRIBE is used by a SIP entity to declare its interest in a particular event. A SIP entity subscribes to a certain event of a class of events. When the subscribed event occurs, NOTIFY requests are sent containing information about the session.

Figure 6-13 shows how the automatic recall service previously described can be implemented using SUBSCRIBE and NOTIFY. Bob calls Laura, but she is busy. Bob sends a SUBSCRIBE to Laura's SIP UA indicating his interest in Laura's status. Laura's SIP UA answers with a 200 OK to the SUBSCRIBE and sends a NOTIFY to Bob with Laura's current status: busy. After a while, Laura hangs up and her UA detects a change in her status and sends a NOTIFY to Bob with her new status: available. Bob will then send an INVITE to Laura to talk to her. This time his call will be successful.

Figure 6-14 shows another example of a service implemented using SUBSCRIBE and NOTIFY. Bob is participating in a SIP conference call that uses a centralized conference unit. All of the participants send an INVITE to the conference unit to establish a session with it. The conference unit then combines all incoming audio streams and delivers them to the participants.

Bob wants to know how many people are attending the conference at any moment. He sends a SUBSCRIBE to the conference unit. The conference unit keeps Bob updated about changes in the number of participants by sending a NOTIFY every time a participant joins or leaves the conference.

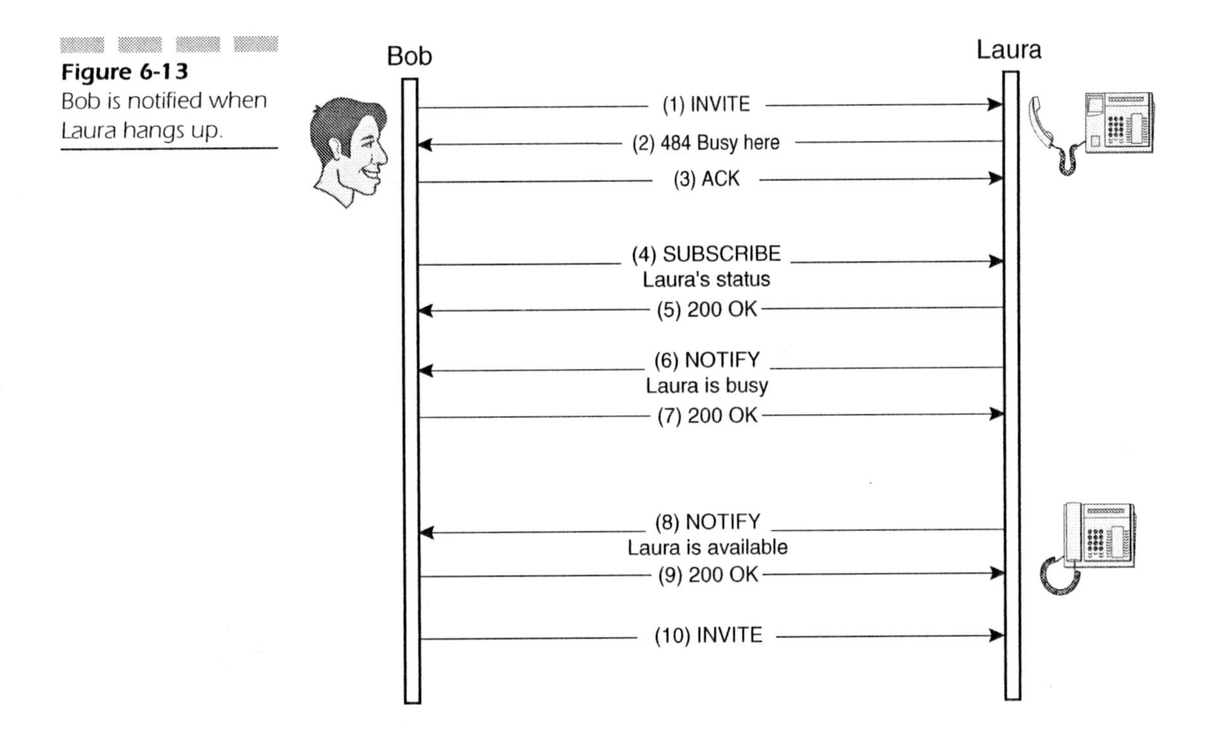

Figure 6-13 Bob is notified when Laura hangs up.

When Laura joins the conference, Bob receives a NOTIFY with the new number of participants.

Third-party Call Control

We know a SIP UA can establish a session with other UAs. However, a user sometimes wants to establish a session between other UAs without taking part on it. This situation is quite common when one of the UAs represents a machine instead of a human. If a conference call is scheduled using a conference unit like the one described in the previous section, invitees can opt out without losing out. Bob is interested in what is said during the conference call, but he is busy at that particular time and can't attend. Bob thinks that recording the conference would be a good idea.

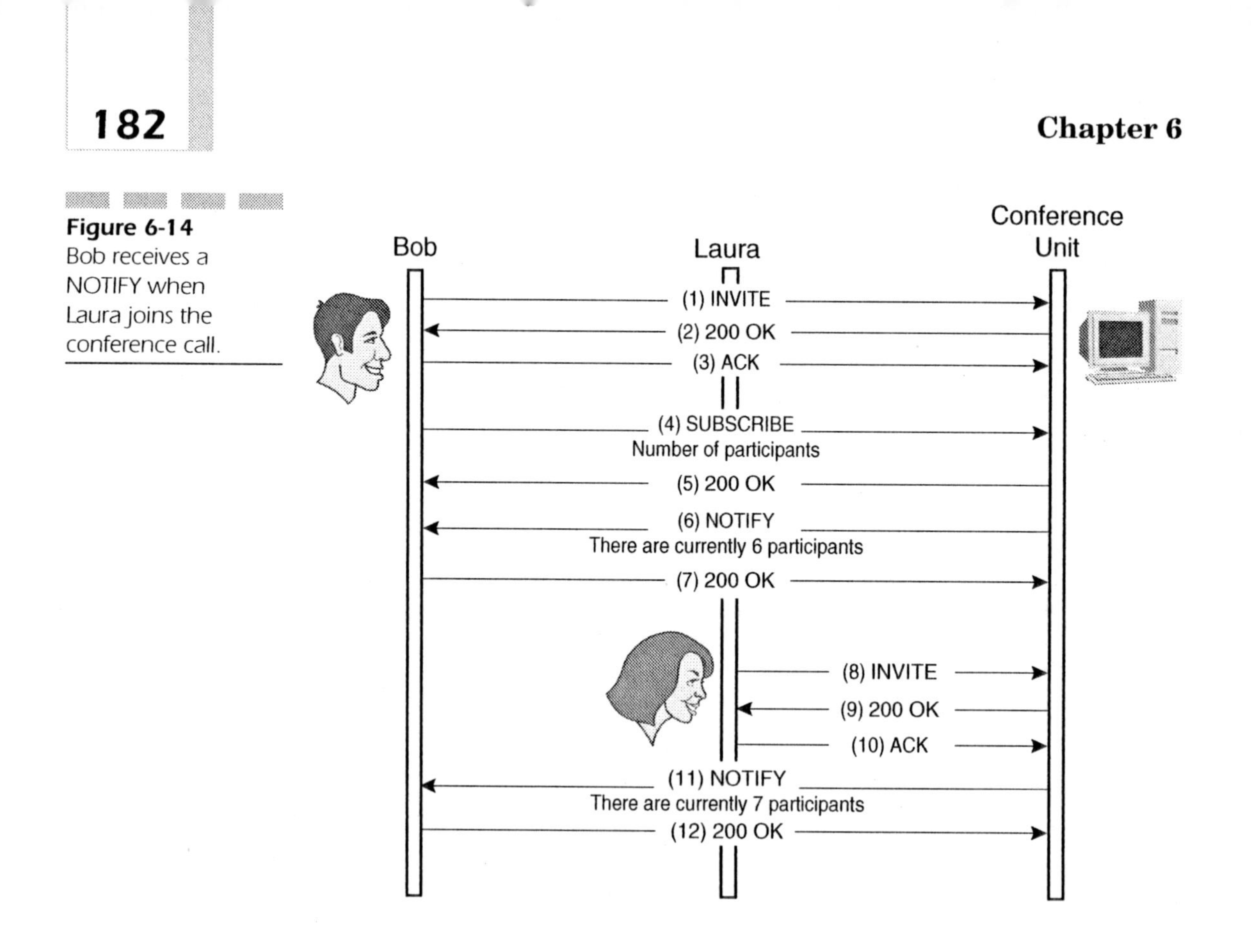

Figure 6-14 Bob receives a NOTIFY when Laura joins the conference call.

However, Bob only has his SIP mobile phone with him, and it has very limited storage capacity. He'll need to establish a session between the conference unit and his home PC. To resolve his problem, Bob resorts to third-party call control, which enables a SIP entity to manage sessions between other parties. A controller that establishes a session between two participants will be involved in the SIP signalling, but not necessarily in the session media. In our example (Figure 6-15), Bob becomes the controller and establishes an audio session between the conference unit and his home desktop.

How It's Done Third-party call control [draft-rosenberg-sip-3pcc] does not require any extension to the core SIP specification. The controller invites the participants and then exchanges the session descriptions between them. Figure 6-16 shows how this example is implemented. Bob sends an INVITE without any session description. Because Bob's computer receives an empty INVITE, it expects that the session description will be sent in the ACK instead. Bob's computer follows normal SIP procedures and returns a session description in a 200 OK response to Bob who then uses

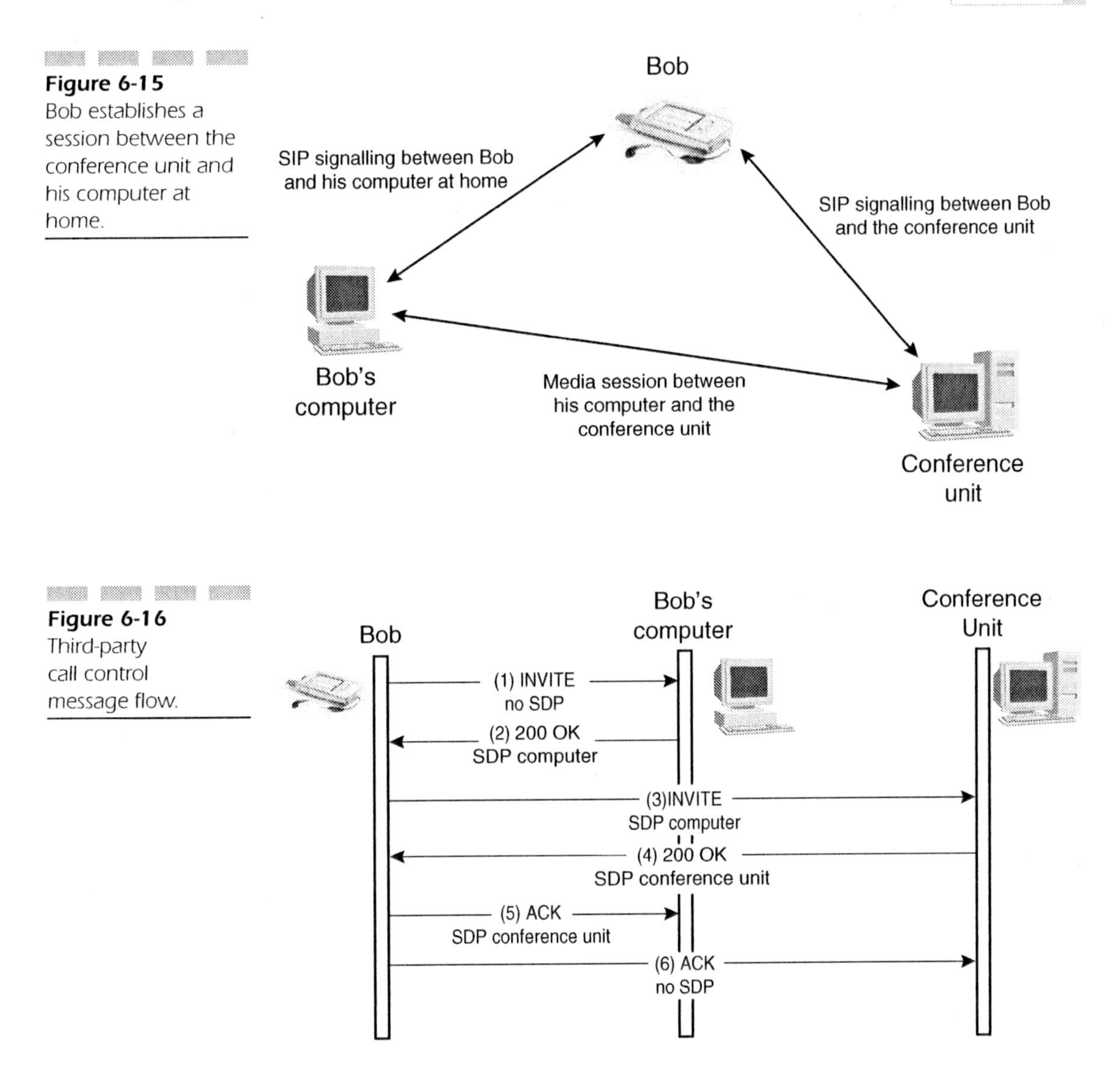

Figure 6-15 Bob establishes a session between the conference unit and his computer at home.

Figure 6-16 Third-party call control message flow.

this session description to INVITE the conference unit. Because this second INVITE contains the session description provided by Bob's computer, the conference unit will send the audio stream to Bob's computer. It responds with a 200 OK. Bob receives the session description of the conference unit in this 200 OK and sends it to his computer in an ACK. Now the media stream flows directly between the conference unit and Bob's computer while SIP signalling traverses Bob's SIP mobile.

The most important feature of this message flow is that it uses nothing but core SIP. Therefore, simple UAs with minimum SIP support can be controlled by a SIP controller.

Session Transfer

We have seen that third-party call control enables a SIP entity to be in control of session signalling while the media is exchanged between other entities. In some situations, the controller won't want to continue monitoring (controlling) the signalling of the session. Instead, he or she will want the other entities to continue the session independently. At this point, the controller needs a mechanism to transfer SIP sessions to another entity.

The most common example of session transfer occurs when I make a phone call and it's answered by the callee's secretary. I tell her the purpose of my call and whom I want to contact. Then the secretary transfers my call to the covering person or voice mailbox. This is an example of session transfer because once the transfer has succeeded, the secretary immediately stops receiving any signalling or media related to that session.

A SIP extension provides session transfer functionality. In our example, it enables the secretary to transfer the call. Furthermore, it enables her to check whether the transfer was successful or not. In case the transfer has failed, the secretary can regain control of the call for a second attempt.

For example, Laura calls Bob's SIP *Universal Resource Locator* (URL), but his secretary picks up the phone. In order to transfer the call, she instructs Laura's UA to INVITE Bob at his current location. Laura's UA INVITEs Bob using the new URL provided and terminates the previous session.

Note that if Laura did not terminate the first call at this juncture, they could have easily established a three-party conference using the same mechanism.

How It's Done To provide session transfer functionality, a new method was defined: REFER [draft-ietf-sip-cc-transfer]. The idea behind this mechanism is that one SIP entity instructs another to perform a certain action. Namely, the REFER method instructs a server to send a specific request to a certain URL. Figure 6-17 illustrates the REFER mechanism. Laura calls Bob at his office and gets his secretary instead. Laura explains that she wants to speak to Bob. The secretary puts Laura on hold by sending a re-INVITE with a session description indicating call hold. Then she

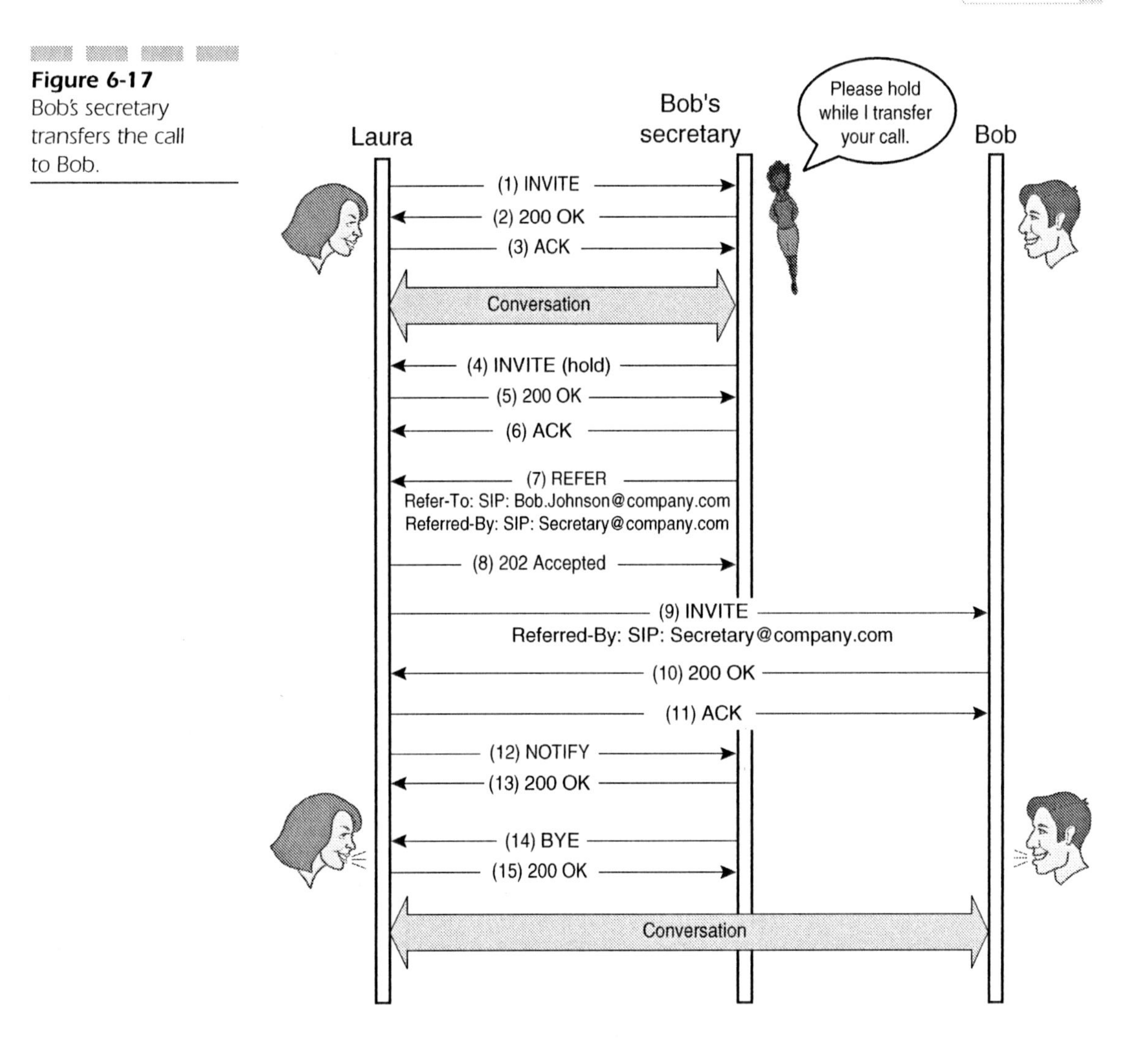

Figure 6-17 Bob's secretary transfers the call to Bob.

sends a REFER to Laura instructing her to send a new INVITE to Bob's current location. The Refer-To header of the request contains the URL where Laura has to send her new INVITE. The Referred-By header contains the secretary's SIP URL. Laura copies the Requested-By header into her new INVITE and sends it to the SIP URL contained in the Request-To header. When Bob receives this INVITE, he knows it is the result of a transfer made by his secretary because the Referred-By header contains her SIP URL.

When the INVITE transaction is completed, Laura sends a NOTIFY indicating the result of the transfer. In our example, the transfer ended successfully, so Bob's secretary sends a BYE to terminate her call with Laura. If the transfer had failed, Bob's secretary would have been able to retrieve her session with Laura to continue speaking with her. After the BYE, Bob and Laura are in a closed two-party session with both signalling and media traffic exchanged directly between them.

Sending Commands

Because SIP doesn't function as a master/slave protocol, it's not suitable for controlling tightly coupled devices. In such scenarios, the master sends commands to the slave who responds with the current status of the command. The master controls how the slave is proceeding with each command at every moment. Protocols such as H.248 or MGCP [RFC 2705] are appropriate to resolve these problems because they were designed with tightly coupled devices in mind.

However, sometimes it is necessary to send a command to a device outside a master/slave architecture. When both the entity issuing the command and the entity receiving it are loosely coupled, SIP becomes an interesting alternative to master/slave protocols. In such an environment, when one entity sends a command to another entity, the latter does what it is requested to do and then reports the status of the command. Note that the entity issuing the command does not behave as a master because it cannot control the process from moment to moment and has to await the final result of the command.

How It's Done A new SIP method called DO [draft-moyer-sip-appliances-framework] was defined to carry commands. An entity issues a DO method carrying a command in the message body. The entity receiving the DO performs the action the message has specified. Besides the DO method, a format to describe commands, the *Device Messaging Protocol* (DMP), was defined. DMP describes commands the same way as SDP describes sessions.

With these two extensions, it is possible to control SIP-enabled entities, and Bob can control his SIP-enabled radio from his laptop. Figure 6-18 shows how Bob's SIP UA adjusts the volume using DO. In this way, the UA can automatically mute the radio every time Bob receives a call and resume the music once the call is over via another DO.

Figure 6-18
Bob controls his radio using SIP.

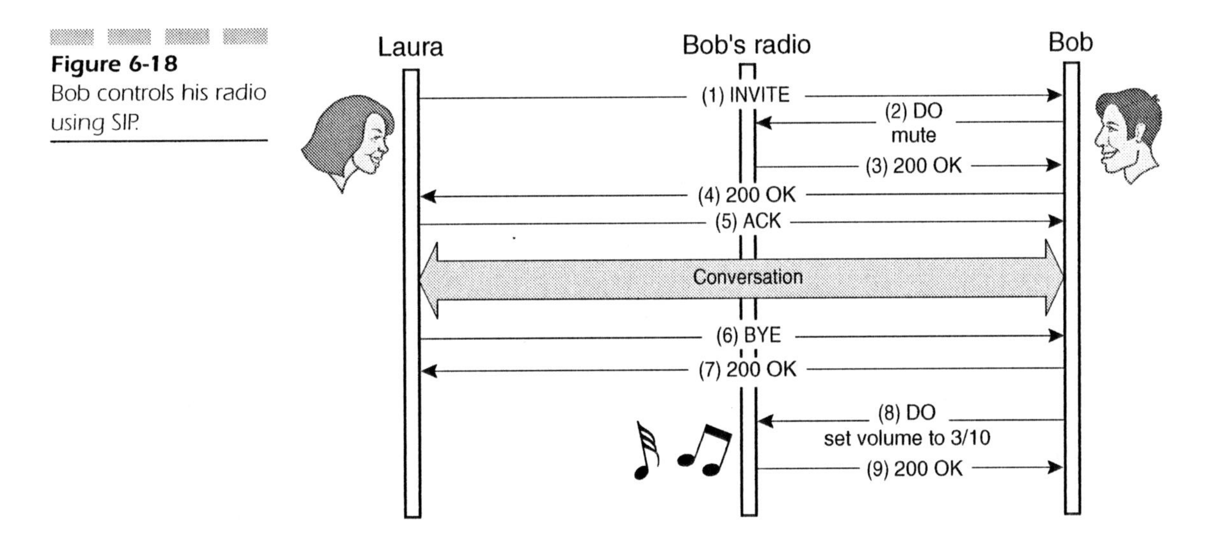

SIP Security

The IETF community considers security a critical aspect of any protocol because the IETF designs protocols for use in the Internet, which is considered a hostile environment. Internet users protect their transmissions in many ways against potential attackers. SIP users are no exception. It is important, however, not to confuse SIP security with the security of a SIP-enabled session. SIP security is concerned with the exchange of SIP signalling. Hence, Bob can send an encrypted INVITE to Laura so that nobody knows which kind of session they are establishing, but once the session is established, if they transmit unencrypted RTP packets, an eavesdropper will be able to hear the whole conversation.

SDP, for instance, can carry cryptographic keys for encrypting media sessions. SIP users can exchange keys during session establishment and then use them to exchange media in a secure way. The remainder of this section focuses on SIP security (authentication, message integrity, and confidentiality) rather than on media security.

How It's Done A major aspect of security is authentication. When Laura receives a SIP request from somebody claiming to be Bob, she wants to be sure that Bob is really the one sending the request. She needs a mechanism to check the identity of any caller.

Because SIP is based on HTTP, it can borrow HTTP authentication mechanisms used in the Web for message authentication based on user IDs and passwords. When somebody tries to view a restricted-access Web page, HTTP has to determine who the user is before it can decide whether to give him or her access. In SIP, these mechanisms are used to accept or reject a session invitation. Let's briefly analyze two different authentication schemes: basic and digest. In both schemes, a server challenges the client with a 400-class response that asks for the appropriate credentials. The client responds with a new request containing what the server required. Challenges are carried in the WWW-authenticate SIP header and credentials are carried in the Authorization header (Figure 6-19).

In the basic authentication scheme, the client provides a user ID and a password as credentials. This mechanism has an important limitation: both user ID and password are sent in clear text. Thus, any eavesdropper can readily obtain them just by sniffing the network.

The digest authentication scheme overcomes that limitation. It is also based on user IDs and passwords, with the difference that they are never sent through the network. The server challenges the client sending a nonce value. Then the client calculates a checksum of the nonce value, the Request-URI, the SIP method, the user, ID and the password and sends it to the server. The server can thereby confirm that the client knows the user ID and password without ever exposing them.

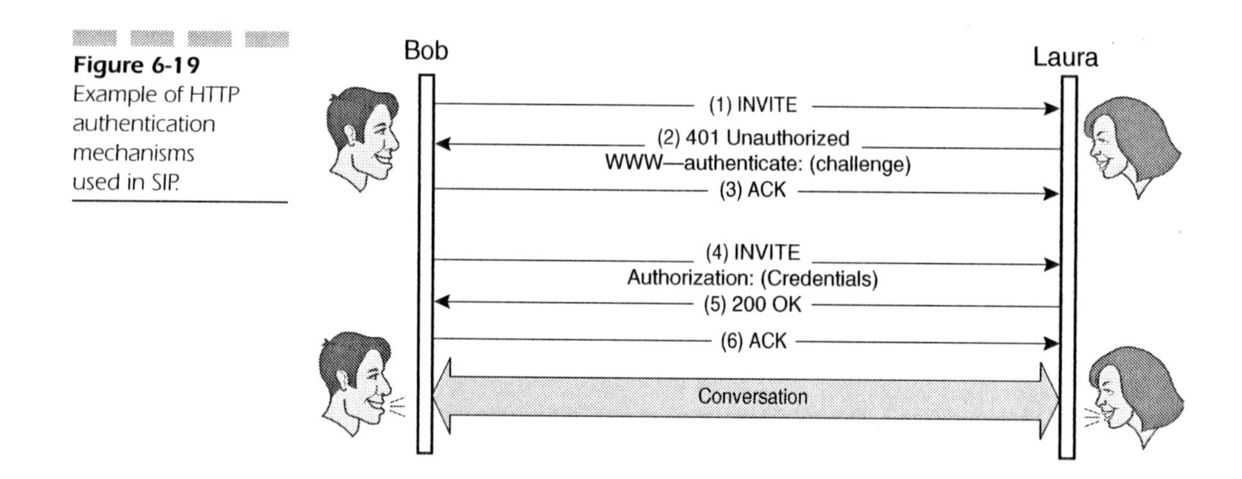

Figure 6-19 Example of HTTP authentication mechanisms used in SIP.

These two mechanisms provide a level of authentication, but are not enough for applications that require strong authentication. The basic and digest schemes have serious limitations. A malicious proxy in the middle of the SIP path could change the contents of the SIP message generated by Bob and forward it to Laura. When Laura checks the Authorization header of the message, she'll believe that it's from Bob. However, the proxy has changed some parameters of Bob's request so that even if Bob did generate the original message, what Laura received is not really what Bob sent.

To overcome this weakness in the basic and digest authentication schemes, it is necessary to provide message integrity together with authentication. This is the only way a server can be sure that the SIP message received was not modified by any entity in the network after being sent by the client.

S/MIME Authentication and Message Integrity A general security mechanism is usable with any transport mechanism that transports MIME [RFC 2045] data. Known as *Secure/Multipurpose Internet Mail Extensions* (S/MIME) [RFC 2633], it is generic enough to exchange secure e-mails and work with both HTTP and SIP. S/MIME provides a format whereby the contents of a message are signed, enabling the receiver to verify its integrity.

Messages are signed using a public-key encryption mechanism. An individual user has two keys: the private key, which only the user knows, and the public key, which is available to anyone. Something encrypted with a private key can only be decrypted with a public key and vice versa.

In order to sign a message, the user calculates a digital signature by using his or her private key and the contents of the message as input. Then it adds the signature to the message. This way, if Bob signs a message using his private key, Laura can check the integrity of the message and verify that it comes from Bob. She uses Bob's public key to check that the digital signature belongs to Bob, who is the only one who knows his private key.

S/MIME Confidentiality However, even if Laura can ascertain that Bob was the sender and that the message was not modified in the network, any eavesdropper can still see the contents of the SIP messages exchanged. S/MIME provides an encrypted-only format to ensure that message contents remain confidential. Bob encrypts the contents of his messages with Laura's public key. Because Laura is the only one who knows her private key, Bob's messages can only be decrypted by her. Both formats provided by S/MIME, signed-only and encrypted-only format, can be combined to provide authentication, message integrity, and confidentiality.

End-to-End and Hop-by-Hop Security We have seen that proxies need to examine certain headers such as Request-URI, Via, To, From, Cseq, and Call-ID in order to route requests and responses properly. Therefore, these headers cannot be encrypted end to end. Instead, it is possible to use hop-by-hop encryption. A secure channel can be established between a UA and a proxy or between two proxies. Everything sent through this secure channel is encrypted. However, a UA using a secure channel towards a SIP proxy cannot be sure that the latter will use another secure channel towards the next hop. End-to-end and hop-by-hop security are complementary and should be used together.

Note, however, that hop-by-hop security falls outside of the scope of SIP because hop-by-hop encryption is typically done at lower layers using *Internet Protocol Security* (IPSec) [RFC 1827] or *Transport Layer Security* (TLS) [RFC 2246]. Secure channels established at a lower layer remain transparent to the application layer. End-to-end encryption of the message body is essential because some session description protocols such as SDP carry keys for encrypting the media. If the message body were not encrypted, these keys would be exposed to any potential eavesdropper.

In this chapter we have gone through a set of SIP extensions. Let's analyze in the next chapter how these extensions are used by different architectures.

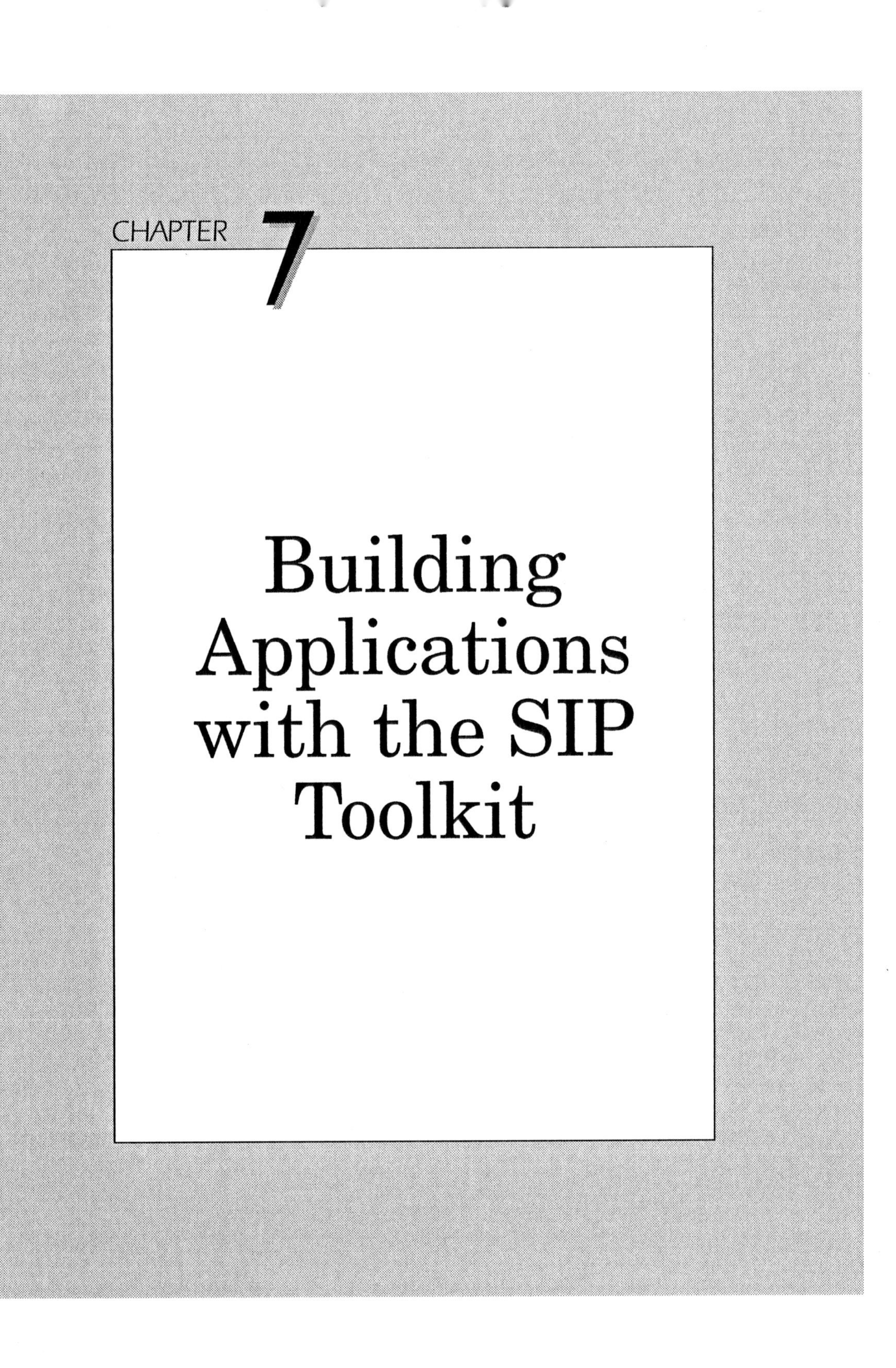

CHAPTER 7

Building Applications with the SIP Toolkit

We have just looked at several extensions to *Session Initiation Protocol* (SIP) that enhance the protocol. Each extension provides a particular functionality that might be useful under certain circumstances. In order to provide a particular service, an application chooses a subset of these extensions and combines them to produce an expected behavior. Therefore, it is the rule and not the exception for different services within different architectures to use different sets of extensions. The general feeling within the SIP community is that the SIP toolkit is now complete enough for a wide range of services. It is true that some extensions are still in Internet draft status, but the set of extensions as a whole is believed to be virtually complete. Minor extensions will certainly be added in the future to cover missing functionality, but the most important extensions have already been defined and are ready to use in any service architecture.

This chapter describes architectures that use SIP, among other protocols, as a signalling protocol. These architectures define different logical boxes and describe how SIP is used between them with a particular set of extensions. In many examples, we will see that a particular logical box in an architecture might act as a SIP proxy without being called one. Some people have wondered if this is done just to confuse us with terminology and I sympathize. The real reason for doing it is that a box can behave like a proxy with respect to SIP, but also have interfaces where protocols other than SIP are used. Therefore, a particular box might be seen as a SIP proxy in one way whereas it might be seen as an LDAP client in another way. The new names defined by an architecture identify a logical box as a whole, taking into account its global functionality rather than just SIP.

This chapter is especially important for designers because it shows how SIP has been used to resolve a variety of real problems. The exploitation of SIP by different architectures is what makes SIP successful. A very well designed protocol with no everyday applications would not be very useful. We will see that SIP has been tapped by such architectures as the *third generation* (3G) of mobile systems and PacketCable—the architectures that will bring SIP terminals to hundreds of millions of users.

Third-generation Mobile Systems

The third generation of mobile systems has aroused a lot of hype and debate in the telecommunications world. It is seen as the technology that will merge the Internet world and the cellular world. It is claimed that 3G will provide ubiquitous access to all successful services provided by the

Internet. *Second-generation* (2G) mobile phones delivering only voice and short message services between users will be replaced by advanced multimedia terminals equipped with Internet access. Combining Internet technologies with widely available cellular access will make the Internet mobile.

The *Third Generation Partnership Project* (3GPP) develops technical specifications for 3G mobile systems based on evolving *Global System for Mobile communications* (GSM) networks. The organizational partners of 3GPP are the *Association of Radio Industries and Businesses* (ARIB), the *China Wireless Telecommunications Standard* (CWTS), the *European Telecommunication Standards Institute* (ETSI), Committee T1, the *Telecommunications Technology Association* (TTA), and the *Telecommunications Technology Committee* (TTC). The specifications elaborated by 3GPP will be used in 3G systems such as *Universal Mobile Telecommunications System* (UMTS).

Whereas 3GPP develops specifications for 3G systems based on GSM networks, 3GPP2 develops technical specifications for 3G systems based on ANSI/TIA/EIA-41 networks. Both projects, 3GPP and 3GPP2, are part of the *International Telecommunication Union* (ITU) through the *International Mobile Telecommunications-2000* (IMT-2000) process.

Network Domains

The 3G network, as defined by 3GPP, is divided into three different domains: the circuit-switched domain, the packet-switched domain, and the IP multimedia domain (commonly known as the IP multimedia subsystem).

The first of these domains employs circuit-switching technology to provide voice and multimedia over circuits. This domain is not dissimilar to 2G systems, where all services are provided through circuit switching.

The packet-switched domain provides IP connectivity to the terminal. A terminal obtains access to the Internet through this domain. Users can surf the Web, send and receive e-mails, and do most of what can be done in the Internet. This domain does not define any particular architecture on top of IP; it can be seen primarily as an access technology. In earlier mobile generations, people used a dial-up connection or an *Internet Services Digital Network* (ISDN) line for this purpose; 3G systems will provide Internet access through the packet-switched domain. What the user does with this access is strictly up to the user.

The third domain is the most important one from our point of view. The IP multimedia domain provides IP-based multimedia services to users employing SIP as the principal signalling protocol. In 3G architectures, multimedia services take on new importance and are separated out for closer examination.

The IP Multimedia Domain The first question that a reader might ask is how the IP multimedia domain differs from the packet-switched domain. A 3G terminal in the packet-switched domain could theoretically use its cellular Internet access to contact the SIP server of the user's choice and get the service he or she wants. However, although the logic behind this scenario is real, the characteristics of cellular access make it unlikely to happen. In order to obtain efficient data transport, a radio interface must be configured with full awareness of the characteristics of the data that will be transported. Otherwise, data transfers slow down and fall short of the requirements for real-time traffic. Voice cannot transmit over a radio interface unless the network knows which codec is used, with what port numbers, IP addresses, and so on. Only then can the network apply techniques such as error protection and header compression to make the voice transfer efficient. These techniques also improve the perceived quality of voice received by a terminal beyond anything that could be delivered over the packet-switched domain. Herein lies the main reason to couple access and multimedia service provision in the cellular environment.

IP Multimedia Architecture SIP is used as the principal signalling protocol within the IP multimedia domain. All 3G terminals contain a SIP *User Agent* (UA), and IP multimedia network nodes consist of SIP proxies. However, as we explained before, these are not actually called SIP proxies. Instead, they're referred to as *Call/Session Control Functions* (CSCFs).

The 3GPP architecture defines three types of CSCFs, each with a different role. They are the *Proxy CSCF* (P-CSCF), *Serving CSCF* (S-CSCF), and *Interrogating CSCF* (I-CSCF). The P-CSCF is the point of contact between the network and the terminal. Both outgoing and incoming SIP messages traverse the P-CSCF. A P-CSCF acts as an outbound SIP proxy because all SIP requests are sent to it regardless of their final destination.

The S-CSCF provides services to the user. When a terminal REGISTERs, it is associated with an S-CSCF, which provides it with services that the user is subscribed to. Both outgoing and incoming sessions will traverse the S-CSCF associated with the terminal. This way, an S-CSCF can provide services on both types of sessions.

The role of the I-CSCF is to find the proper S-CSCF for a particular user. When an I-CSCF receives a request, it routes to the S-CSCF that has to handle the session. For incoming sessions, the I-CSCF is the point of contact within a provider's network; it receives requests for a user within its domain and routes them to the proper S-CSCF. For outgoing sessions, the I-CSCF receives requests from the user and routes them to the associated S-CSCF.

The IP multimedia domain of a provider's network contains many CSCFs of each of these three types.

Another vital node in the IP multimedia domain, albeit not a SIP node, is the *Home Subscriber Server* (HSS). The HSS contains the S-CSCF associated with an individual user and his or her profile. As a result, it knows where the user is reachable and which services he or she is subscribed to. CSCFs consult the HSS when they need such information.

Call Flow Examples

The following examples will help us understand the role of the different nodes in this architecture.

User Registration Figure 7-1 shows terminal registration in a 3G network. The point of such a registration is to assign an S-CSCF to the user. The S-CSCF will be in charge of providing services to the user. This REGISTER from the terminal is sent, as any request from the terminal would

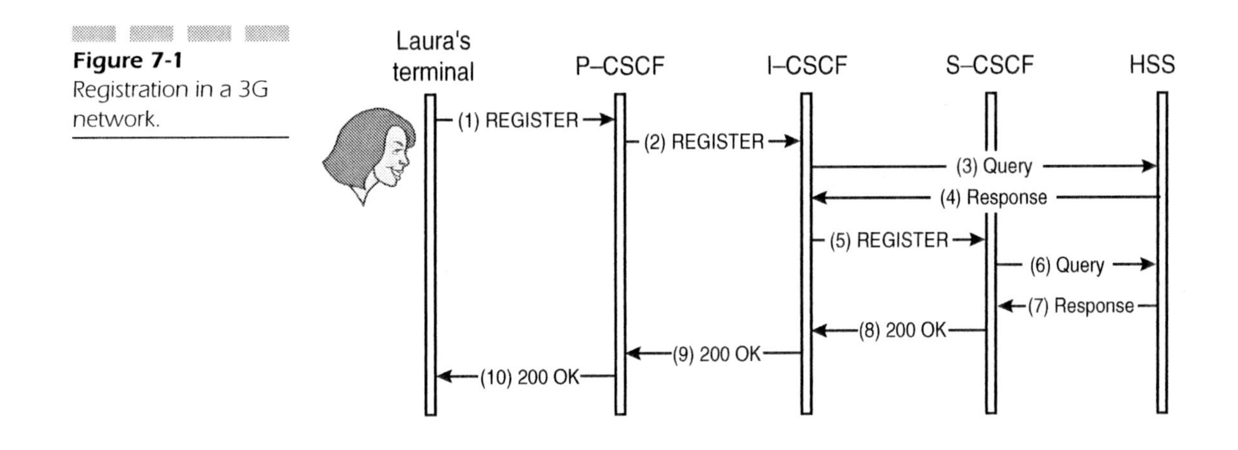

Figure 7-1 Registration in a 3G network.

be, to the P-CSCF. The P-CSCF is the point of contact between the terminal and the network. The P-CSCF sends the REGISTER to the I-CSCF. The I-CSCF has to choose an S-CSCF for the user. To do that, the I-CSCF consults the HSS. Note that the protocol used between them is not SIP. The HSS can be thought of as a location server. It interacts with SIP entities, but does not speak SIP.

The response from the HSS contains enough information so that the I-CSCF can select a particular S-CSCF for the user. Then the REGISTER is forwarded to the chosen S-CSCF. Upon receipt of the REGISTER, the S-CSCF downloads a user profile from the HSS, telling it which services the user requires. Finally, it answers the REGISTER with a 200 OK response.

Registration from a Visited Domain The previous scenario becomes more complex when the user is not on his or her home network. A central requirement for 3G systems is that users must be able to roam from the network of their service provider to any other network (global roaming). This way, a user can tap into the services to which he or she is subscribed even if he or she is on vacation or traveling internationally. Whenever a user is roaming, two distinct domains come into play: the home domain and visited domain. The former is the IP multimedia domain of his or her service provider and the latter is the IP multimedia domain of the network that he or she is currently using.

When a user establishes a session, his or her SIP messages typically traverse a set of CSCFs in both the visited and the home domains. Figure 7-2 shows how a user sends a REGISTER from his or her visited domain to his or her home domain. The message flow is exactly the same as in the previous picture, but in this example, it is possible to distinguish both domains.

When the P-CSCF of the visited domain receives the REGISTER, it detects that this user is not at his or her home domain. The P-CSCF forwards the REGISTER to the user's home domain. From that point on, the process is the same as in any registration from the home domain. The I-CSCF chooses an S-CSCF by consulting the HSS and the S-CSCF downloads the profile of the user.

Note that after the registration, the HSS knows which S-CSCF has been associated with the user. Therefore, in case of an incoming INVITE for that user, the HSS will be able to provide enough information to locate the S-CSCF in charge of the user. The S-CSCF will then forward the INVITE to the user. This scenario is described in the following example.

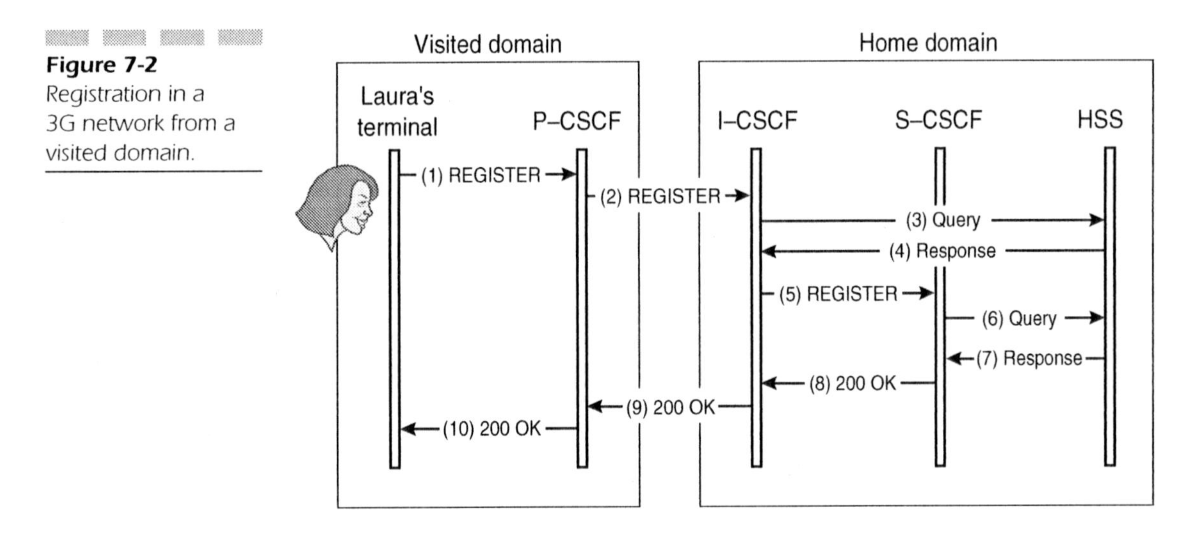

Figure 7-2 Registration in a 3G network from a visited domain.

Session Between Users Who Are Roaming Figure 7-3 is the message flow of a session establishment between two users who are roaming. This is the most general scenario because it shows an outgoing and incoming session from a visited domain.

Let's parse the roles of the different network nodes in this session established between Bob and Laura. Bob is on a business trip to another country. When he arrives at his destination, he REGISTERs his current position as in Figure 7-2. Laura is on vacation and also away from her home domain.

Laura sends an INVITE to the P-CSCF. The P-CSCF notices that Laura belongs to another provider and forwards the INVITE to the proper network. The I-CSCF of Laura's home domain consults the HSS (not shown in the picture) and sends the INVITE to the S-CSCF that was assigned to Laura when she registered. The S-CSCF will send the INVITE to Bob's home domain.

From Bob's home domain's point of view, this is an incoming session for a user who is roaming. The I-CSCF consults the HSS (not shown in the picture) and forwards the INVITE to the S-CSCF that handles Bob. The S-CSCF knows that Bob is currently in a visited domain because he previously registered his current location. Therefore, the INVITE is forwarded to the P-CSCF near Bob, where Bob finally receives it.

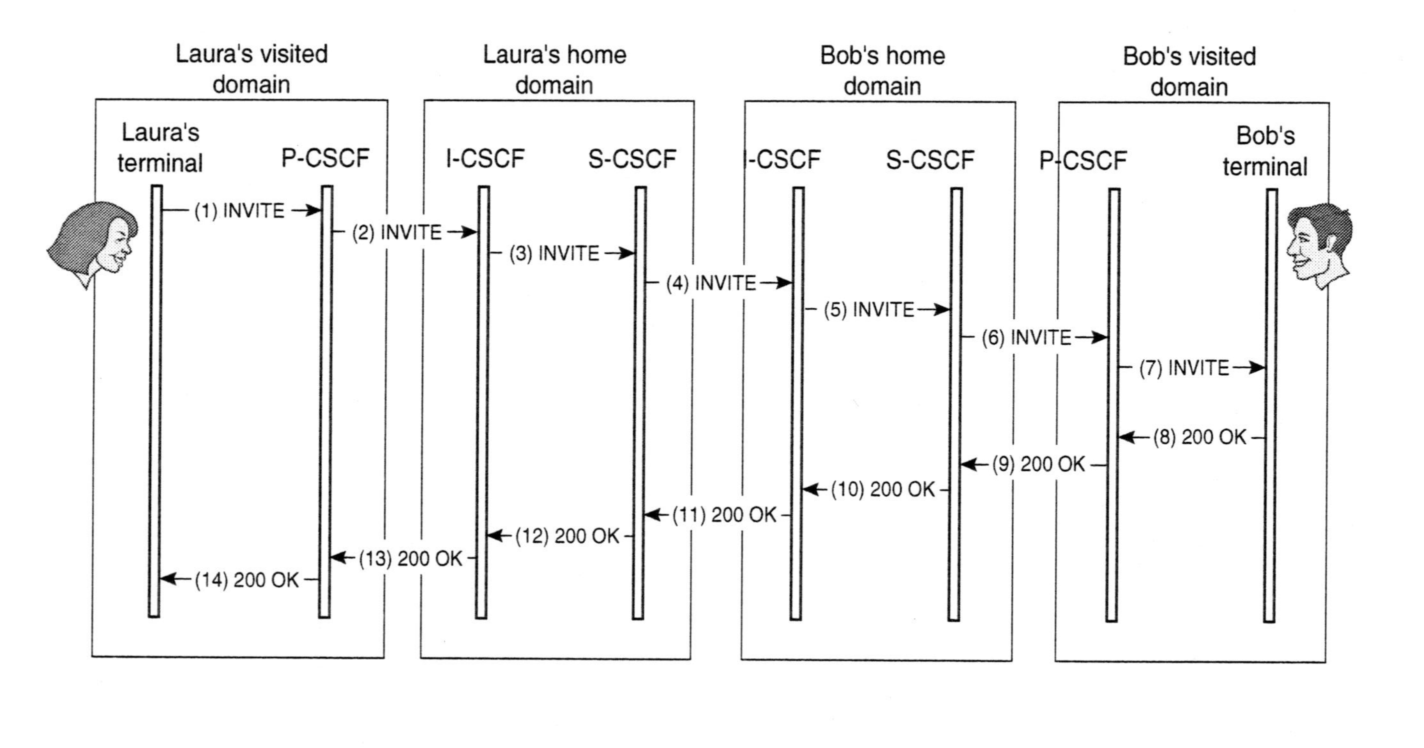

Figure 7-3 Session between roaming users.

Instant Messages and Presence

Instant messages and presence information are great examples of using the SIP toolkit to provide a combined service. We saw in previous sections how instant messages can be sent using the MESSAGE request. We also saw how a user can subscribe to certain events using the SUBSCRIBE and the NOTIFY methods. An application combining these two SIP extensions can provide both instant messages and presence information.

The service in question provides a user with information about the status of his or her friends and enables the user to send them instant messages. SIP is another way to establish multimedia sessions by using the same network's infrastructure of SIP servers.

The user interface of such an application is usually a *buddy list* displayed on the screen. Each person in the list has a label indicating his or her current status. Because Laura has included Bob in her buddy list, his name will appear on the list. If Bob is busy on his workstation, his name appears with the label *available*. When Bob goes for lunch, his label changes to *unavailable*. This way, Laura knows if she can send him a short message at any given moment.

Now Laura clicks on Bob's name, selects *send instant message,* and dashes a message off. If, after exchanging some messages, Laura wants to make a voice call to Bob to clarify something, she just has to click again on Bob's name and select *establishing a voice call* instead.

SIMPLE Working Group

The *Instant Messaging and Presence Protocol* (IMPP) working group was chartered to build a protocol that enables applications combining instant messages and presence information. However, after working on the requirements for such a protocol [RFC 2779], the working group could not come to a consensus. Different groups of people within the IMPP working group had completely different proposals. All of the different proposals had pros and cons.

After merging some proposals, the working group ended up with three main approaches to provide instant messaging and presence service: *APplication EXchange* (APEX), *Presence and Instant Messaging* (PRIM), and *SIP for Instant Messaging and Presence Leveraging Extensions* (SIMPLE).

We will focus on the third approach, SIMPLE, because it is based on SIP. The SIMPLE working group met for the first time at the 50th *Internet Engineering Task Force* (IETF) meeting in Minneapolis (March 2001). Its task was to define a set of extensions that would let SIP provide instant messaging and presence service. Using SIP with these extensions will meet the requirements outlined by the IMPP working group for every IETF protocol that they might conceivably adopt for this purpose. When various approaches to a problem are under consideration, the minimum requirement is that all protocols developed for instant messaging and presence interoperate [draft-ietf-impp-cpim].

Presence Architecture

Basically, an instant messaging and presence service is implemented using two SIP extensions: the MESSAGE method and the SUBSCRIBE/MODIFY framework. The architecture used for presence consists basically of two nodes: the *Presence User Agent* (PUA) and a presence server.

The Presence User Agent, as we call a SIP UA used for presence services, REGISTERs the user's status with the presence server. Hence when Bob logs off his laptop, Bob's PUA will send a REGISTER reporting that Bob is currently unavailable to the presence server.

Remember that a single user might have several PUAs. For instance, Bob has a PUA in his workstation and another one in his laptop. Thus, the presence server gathers REGISTERs from these two devices to keep track of a single user.

Suppose another user is interested in Bob's presence information, so he subscribes to Bob's presence server. After that, every time Bob's presence information changes, the presence server will transmit a NOTIFY request with new status information. Laura is usually interested in what Bob is up to and her PUA receives a NOTIFY saying that Bob has logged off. Laura will see the label on her screen next to Bob's name change from available to unavailable.

In Figure 7-4, Bob's two PUAs send presence information through REGISTERs to the presence server. Laura SUBSCRIBEs to Bob's presence information for updates on Bob's presence.

This example shows how Bob's PUAs send a simple form of presence information: available or unavailable. However, SIP can furnish much more information than this. Bob might be unwilling to receive voice calls, but

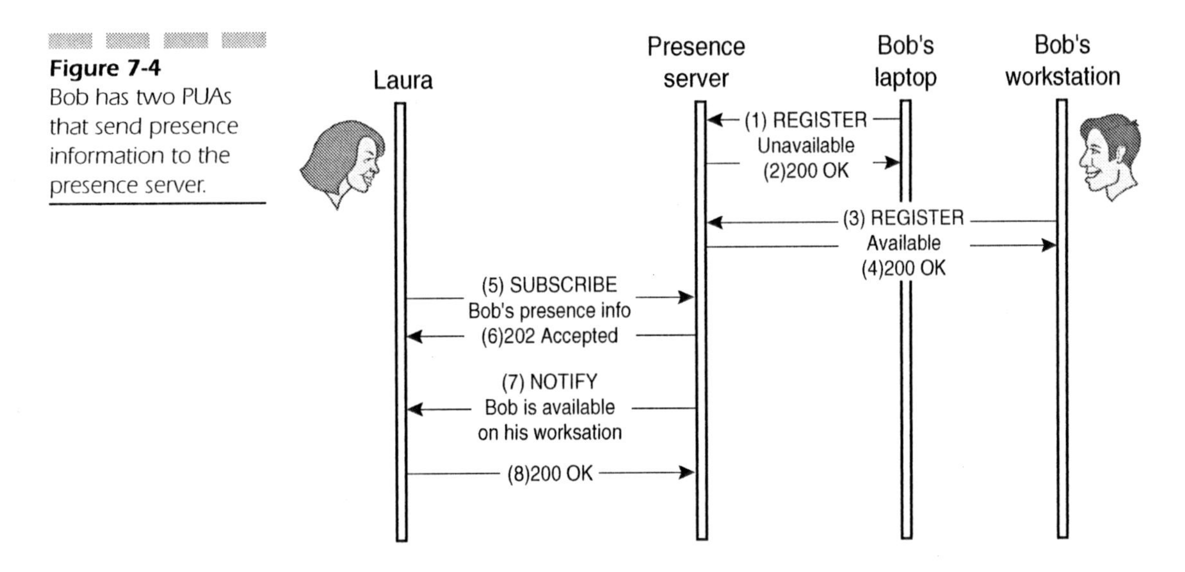

Figure 7-4 Bob has two PUAs that send presence information to the presence server.

want to receive instant messages. Or Bob could even be unwilling to receive voice calls from his boss, but happy to receive them from Laura. The SIP REGISTER method and the SUBSCRIBE/NOTIFY framework provide SIP with enough flexibility to implement this type of service in some granularity.

Instant Messaging

Although presence information can be used to provide a wide range of services, instant messaging is the one that has been most closely associated with presence. Simple presence systems that just provide available or unavailable status have typically been used to tell users who is or isn't receiving messages.

We've stipulated that the MESSAGE [draft-rosenberg-impp-im] method can be used to send instant messages in SIP (Figure 7-5). The advantage of using SIP is that it does more than just enable presence information and instant messaging to be combined. SIP also lets presence information be utilized in establishing any kind of session, including instant messaging and multimedia sessions. Therefore, for applications that offer combined services, SIP is definitively the choice as a presence and instant messaging protocol.

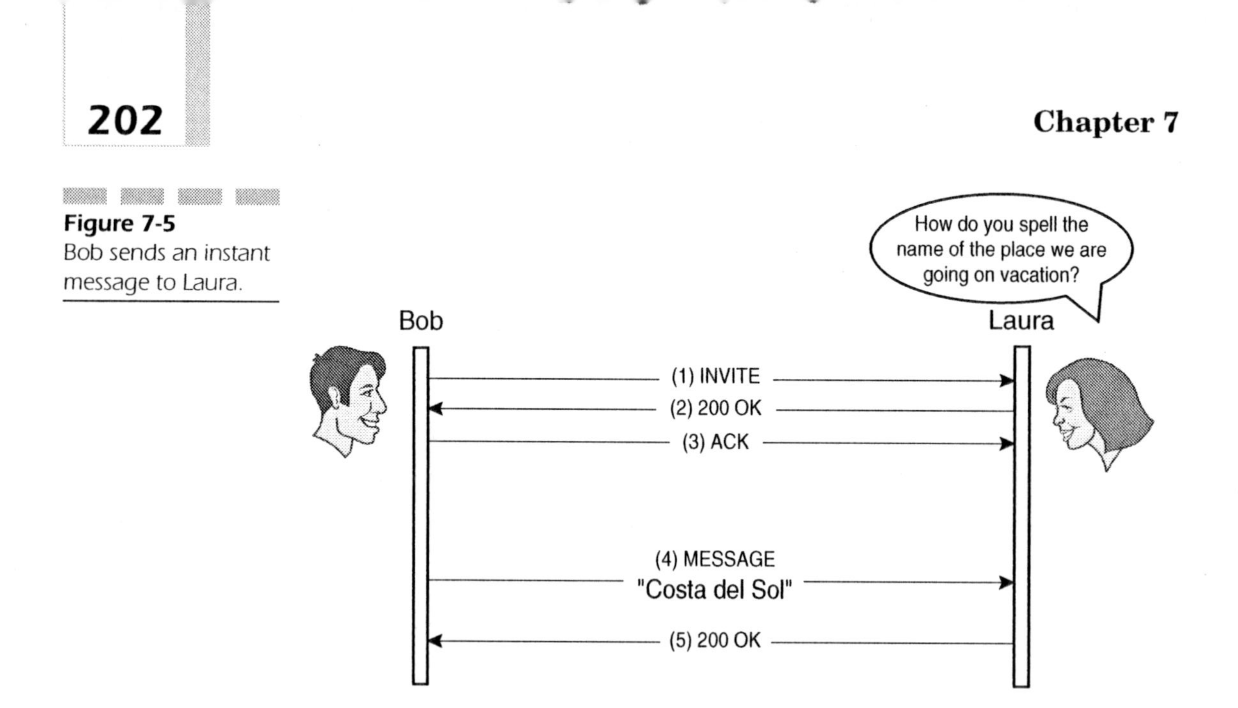

Figure 7-5 Bob sends an instant message to Laura.

PacketCable

PacketCable is a project carried out by Cable Television Laboratories and its member companies. Its purpose is to offer audio, video, and multimedia services through the cable access network. It's envisioned that cable TV subscribers will be able to use their cable modem to make phone calls, for instance. The concept is to provide combined services through the same access technology.

This project represents a big endorsement for SIP. As a 3G network, PacketCable addresses real customer needs using SIP with some extensions. This type of project is a critical source of feedback and new ideas for the SIP community. New requirements lead to the implementation of new extensions.

Cable modems are everywhere in much the same way phones are, and the PacketCable project will provide SIP multimedia services to a much larger number of users.

Architecture

The PacketCable architecture defines a set of nodes and the protocols that will be used between them. SIP is one of the protocols used in this architecture, augmented by a set of extensions. The most important nodes related to SIP operation in the PacketCable architecture are the *Multimedia Terminal Adaptor* (MTA) and the *Call Management Server* (CMS).

The MTA resides on user premises and has a user interface that resembles a traditional telephone (POTS). Between the CMS and an MTA, a protocol called the *Network Call Signalling* (NCS) protocol comes into play. NCS is an extended variant of the master/slave IETF protocol *Media Gateway Control Protocol* (MGCP) [RFC 2705]. The MTA acts as the slave, sending events and responding to commands issued by the CMS.

For instance, the MTA informs the CMS when the user goes off hook or when he or she types some digits. The CMS can then order the MTA to alert the user or to open a media channel for voice. PacketCable architecture will probably evolve so that the protocol used between an MTA and the CMS is SIP instead of NCS.

The CMS is basically a SIP server. One CMS handles a domain and thus typically controls several MTAs, as shown in Figure 7-6. The protocol used between CMSs is the *Call Management Server Signalling* (CMSS) protocol. This is basically SIP with some extensions such as the reliable delivery of provisional responses and *Quality of Service* (QoS) preconditions.

Call Flow Example

Figure 7-7 illustrates SIP operation between two CMSs. Observe how the different SIP extensions implemented in PacketCable systems work together in a call. PacketCable implements the reliable delivery of provisional responses and QoS preconditions so that when the callee picks up the phone, the network will always have enough available resources to meet call requirements.

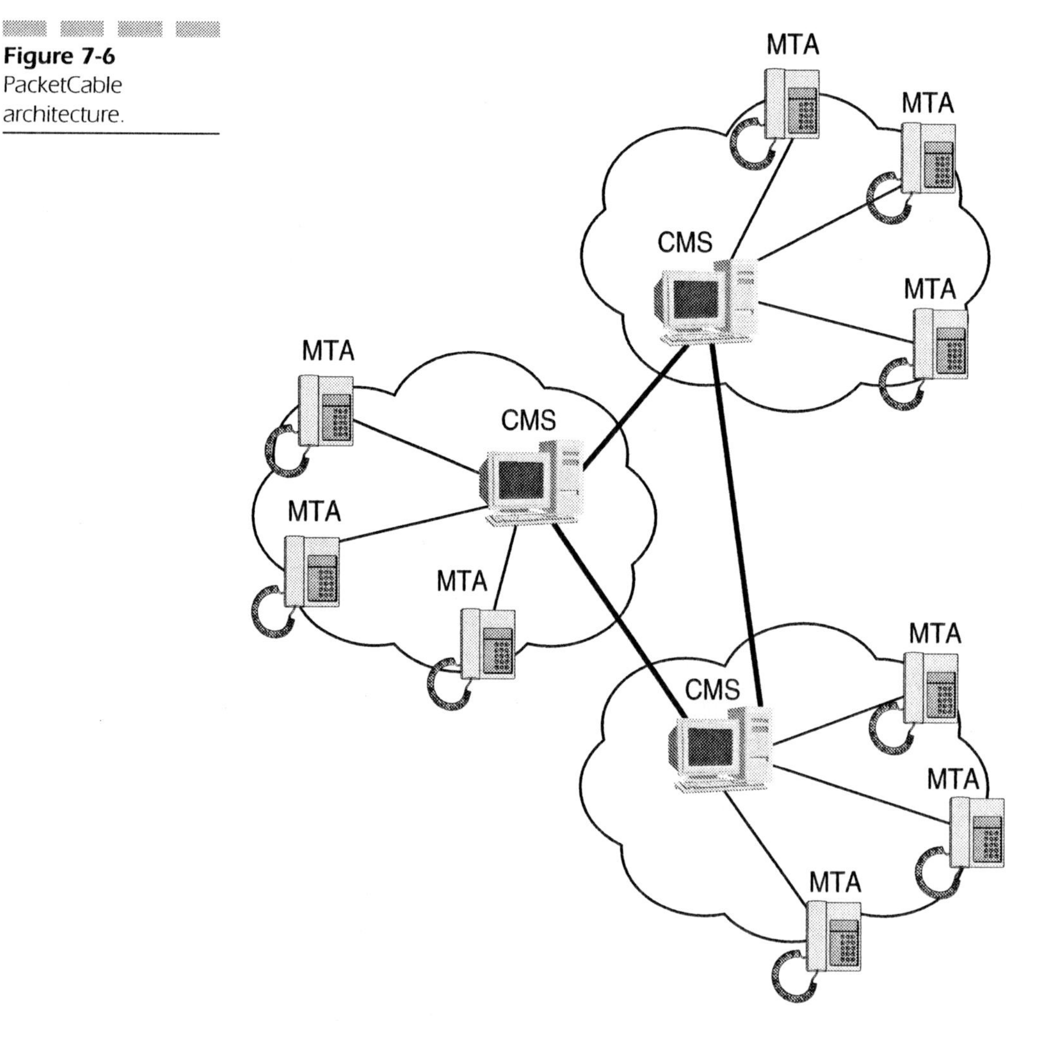

Figure 7-6 PacketCable architecture.

PSTN-to-SIP Interworking

Telephony applications are among the most promising for taking SIP mainstream. In this arena, SIP is clearly an enabler for service builds. Traditional telephony providers are now building IP networks for voice transmission in order to exploit *Voice over IP's* (VoIP's) service flexibility.

When the IP network is properly configured and the appropriate QoS measures are taken, VoIP provides excellent voice quality that is comparable (or even better) than that achieved by the *public-switched telephone net-*

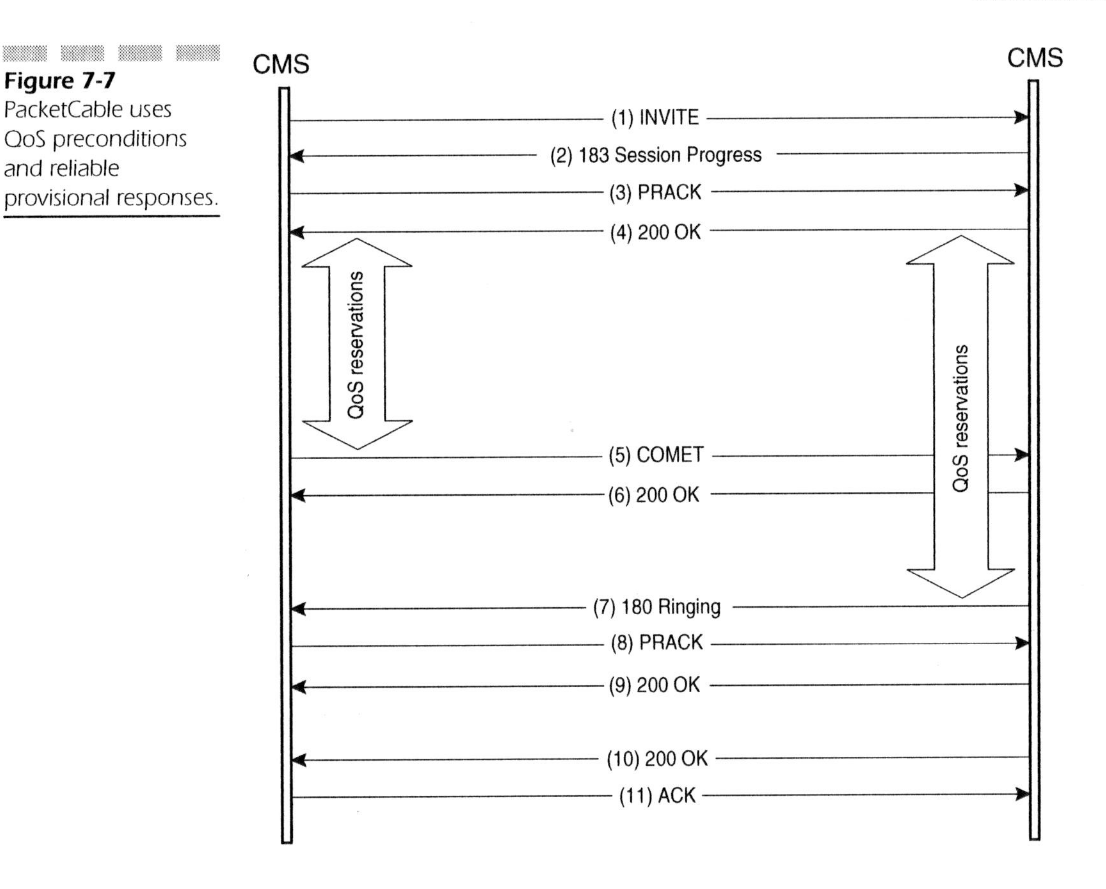

Figure 7-7 PacketCable uses QoS preconditions and reliable provisional responses.

work (PSTN). At present, many VoIP providers offer cheaper international calls than the PSTN. Therefore, it seems inevitable that VoIP will be the future of the telephone network.

However, no matter how efficiently SIP implements telephony services and no matter what level of quality it achieves, a hard-and-fast requirement must be fulfilled by any new telephone system: it must able to interoperate with the PSTN.

The PSTN is the largest telephone network in the world and even if it eventually gives way to IP networks in the future, it will still exist for many years. A huge telephony infrastructure cannot be thrown out and replaced wholesale by new technology.

Given the prominence of VoIP as a SIP driver, the community quickly realized that compatibility with the protocols used in the PSTN was the primary job in making SIP viable. But as we've established in Chapter 1, both the architecture and the design philosophy of the PSTN are at odds with the IETF in general and SIP in particular. Compatibility can only be achieved by implementing gateways to perform protocol conversions in the border between the PSTN and the IP network (Figure 7-8).

Gateways work by acting as a network node for the PSTN and as an end point for the SIP network. With this mechanism in place, the SIP architecture is not influenced at all by the interworking with PSTN protocols. A gateway typically looks just like a SIP UA to the SIP network. Thus, the network responds to calls originated in the PSTN, which are calls originated by another SIP entity as far as SIP is concerned. Gateways let SIP preserve all of its good features even as it connects to the PSTN. In fact, a general rule for SIP interworking with other systems is to ensure that SIP does not have to change behaviors to do so. It is always better to build a complex gateway on the edge for protocol conversion than to push further complexity into the protocols themselves.

The PSTN uses several different signalling protocols and thus several types of gateways exist between the PSTN and SIP.

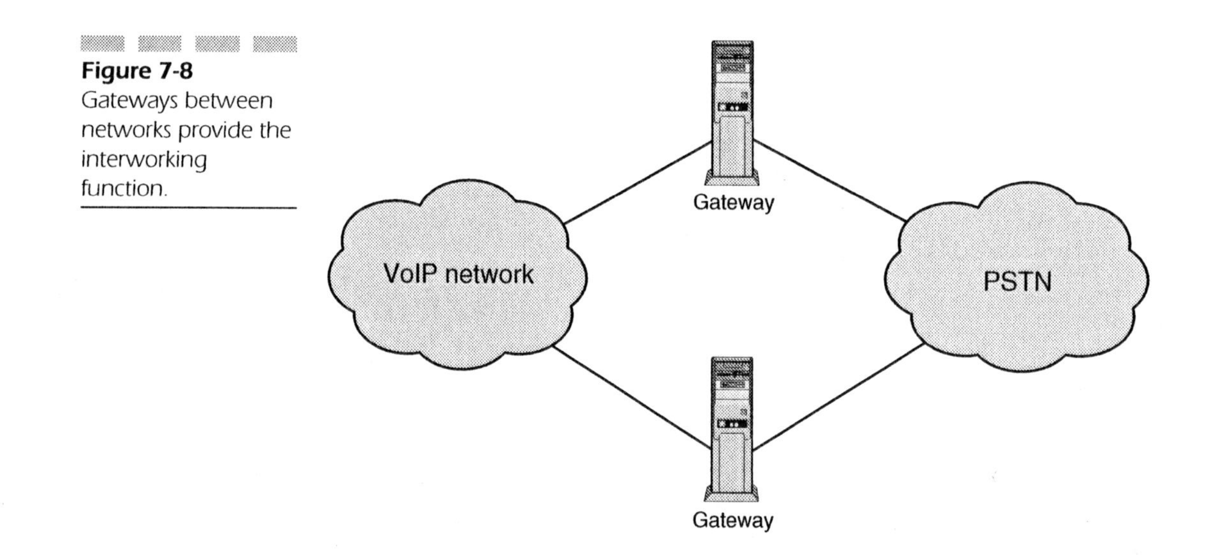

Figure 7-8 Gateways between networks provide the interworking function.

Low-Capacity Gateways

The proper architecture for a particular gateway is a function of its capacity and its requirements. Low-capacity gateways usually integrate signalling and media handling in a single box. They usually interact with PSTN access protocols such as *Digital Subscriber Line No. 1* (DSS-1), and they are intended to support residential or small office requirements.

For instance, a few years ago, Bob had an ISDN phone at home that used DSS-1 to communicate to its PSTN local exchange (Figure 7-9). As technology improved, he decided to replace his old ISDN phone with a small gateway that supports SIP and DSS-1 (Figure 7-10). Today he has connected his new gateway to the small IP network (a LAN) that he has installed at

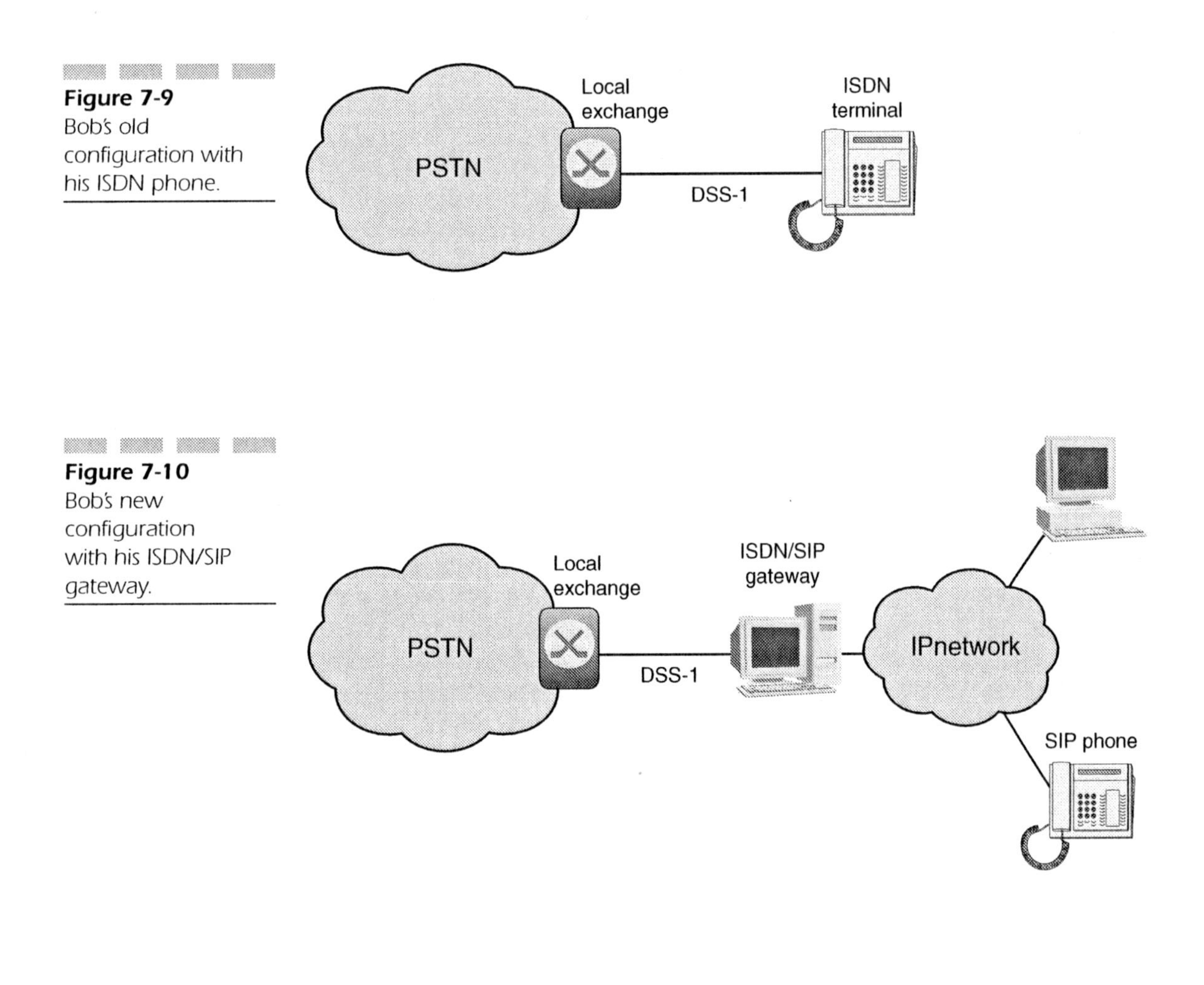

Figure 7-9 Bob's old configuration with his ISDN phone.

Figure 7-10 Bob's new configuration with his ISDN/SIP gateway.

home. The payoff is that Bob can now receive and make calls from his computer or from his SIP phone when he is at home. He has upgraded his residential communications to bring them in line with the business communications he's come to rely on—all without much hassle at all.

Let's look at another case. Laura's office has several phones connected to a *Private Branch Exchange* (PBX) (Figure 7-11). Her boss wants to switch telephone service providers because a new provider is offering more services for cheaper rates. The problem is that all of the employees in Laura's company are used to working with their traditional phones (which, incidentally, they like just fine) and don't want to migrate to SIP phones. To alleviate the need for users to change their behavior and management to invest in new technology, the new service provider installed a gateway, and now company staffers are using a SIP network without really noticing the change (Figure 7-12). They use the same phones, dial numbers in the same way, and receive calls just as before.

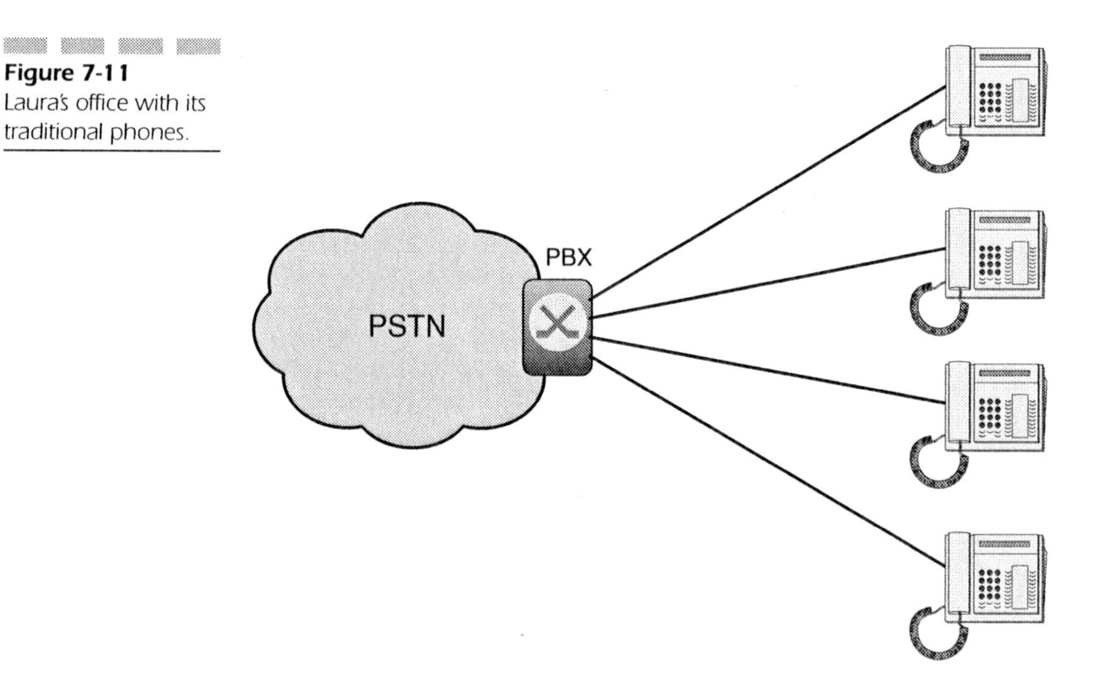

Figure 7-11 Laura's office with its traditional phones.

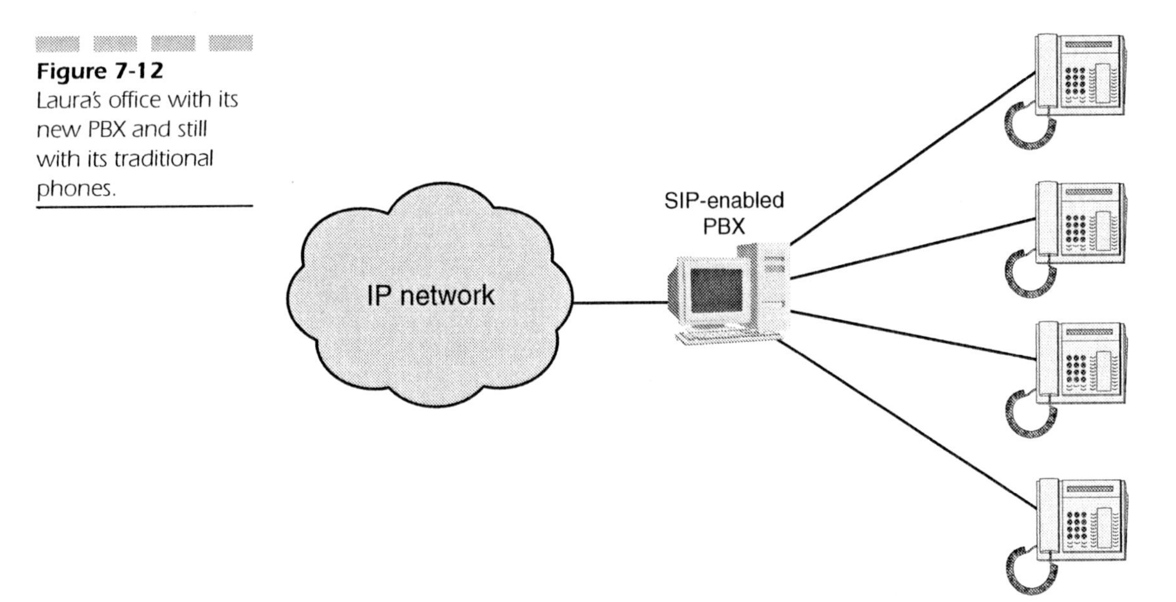

Figure 7-12 Laura's office with its new PBX and still with its traditional phones.

High-Capacity Gateways

As opposed to single-box, low-capacity gateways, high-capacity gateways are usually distributed. That is, different parts of the gateway performing different functions are kept separate. Figure 7-13 shows the most common architecture for a high-capacity gateway:

The *Signalling Gateway* (SG) receives signalling from the PSTN side and encapsulates it over IP (and vice versa). The transport protocol used for this purpose is usually the *Stream Control Transmission Protocol* (SCTP). The SG does not modify these signalling messages; it just routes them to the proper *Media Gateway Controller* (MGC).

An MGC performs two tasks: (1) it converts the signal between the PSTN's signalling protocol (usually *ISDN User Part* (ISUP)) and SIP, and (2) it controls the *Media Gateway* (MG). The MG is responsible for media conversion. It incorporates both voice-over-circuits and VoIP interfaces. An MGC issues commands to the MG such as *open voice channel* or *close voice channel*. The protocol they use to communicate is a master/slave protocol such as MGCP or H.248.

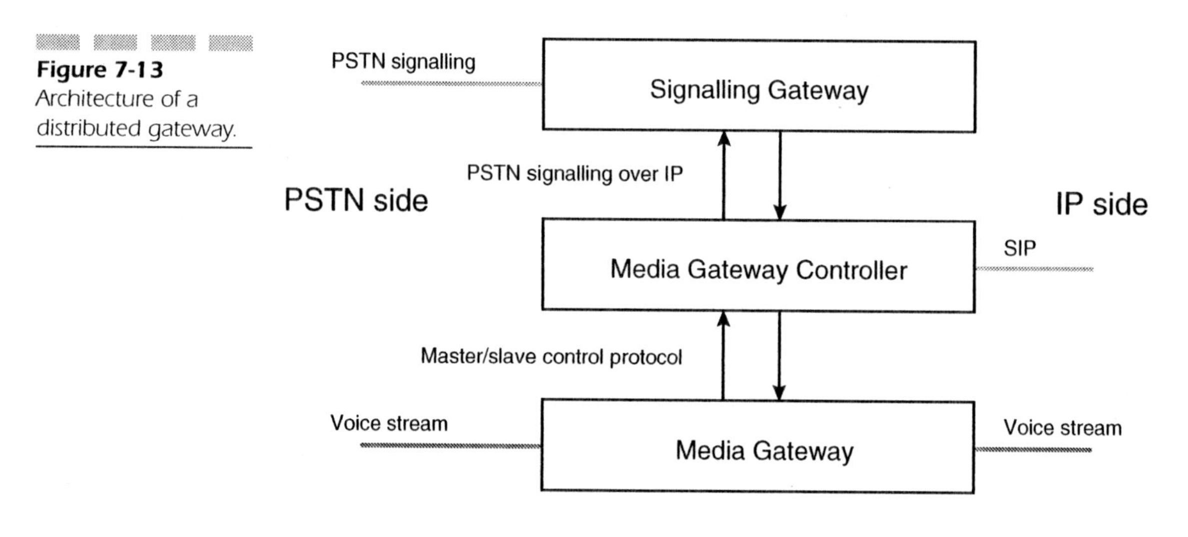

Figure 7-13 Architecture of a distributed gateway.

SIP Extensions for PSTN Interworking

Gateways between SIP and the PSTN normally make use of some common SIP extensions. They usually support reliable delivery of provisional responses in order to deliver progress reports to the PSTN side. Gateways often support QoS preconditions as well. Besides these common extensions, two other extensions were designed with PSTN interworking in mind: the INFO method and MIME media types for ISUP and QSIG Objects. They are used between two gateways for calls that originate in the PSTN, traverse a SIP network, and end up back in the PSTN again. This scenario is referred to as *SIP bridging* (Figure 7-14).

A call between two PSTN phones might traverse a SIP network for many reasons. It might have to utilize some enabling services that are only present in the SIP network. If it couldn't traverse the SIP network, all such services would be off-limits to PSTN users. Another call might be routed through SIP because it's cheaper for the provider. A provider might also use his or her IP network to offload his or her circuit-switched network. Under some circumstances, a call originally destined for a SIP phone can be redirected to a PSTN phone.

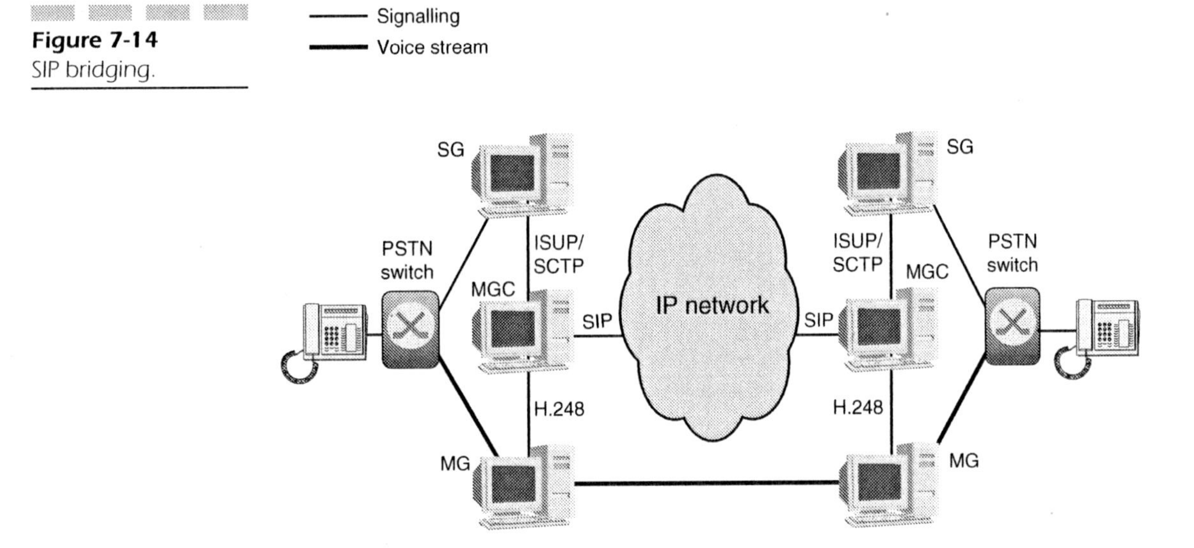

Figure 7-14
SIP bridging.

In SIP bridging scenarios, neither PSTN end user is willing to lose any special feature provided by the PSTN just because the call is traversing a SIP network. Therefore, SIP bridging has to provide a certain level of feature transparency. The information contained in the PSTN signalling messages has to be preserved when converted to SIP messages in the IP network, so that the egress gateway can regenerate the PSTN messages properly at the remote-user terminal.

Multipart Message Bodies Problem: PSTN protocols carry some information that cannot be mapped to any SIP header. Thus, the obvious solution of performing a simple mapping from ISUP to SIP won't work because it leads to some loss of information in the ISUP message. If the information lost happens to relate to a certain PSTN feature, the feature is lost as well.

To prevent information loss, PSTN protocol messages are carried by SIP in message bodies. The MIME media types for ISUP and QSIG Objects were defined for this purpose [draft-ietf-sip-isup-mime]. QSIG is a PSTN signalling protocol used by PBXs. A particular SIP message carries a multipart message body that contains an ISUP message, for instance, and a session description using *Session Description Protocol* (SDP). For the concrete case of ISUP to SIP interworking (which is probably one of the most interesting scenarios), a set of guidelines [draft-ietf-sip-isup] help implementers do a consistent mapping between both protocols.

INFO Method MIME media types solve the problems of sending information in SIP messages that does not map to any SIP header. That's not the whole problem, however. Some PSTN protocols have messages that don't map to any SIP message. For instance, if a gateway receives a PSTN signalling message in the middle of a call, no SIP message can be sent to the egress gateway in order to transfer it from the PSTN. INFO [RFC 2976] method was created for this purpose. The INFO method is used to carry mid-call PSTN messages in its body. Figure 7-15 shows MIME media types and INFO in an interworking situation.

INVITE and the 200 OK carry in their bodies, apart from the session description, a PSTN signalling message. ACK might carry another PSTN signalling message, but it typically does not because PSTN signalling protocols rarely use a three-way handshake.

Once a session is established, the ingress gateway receives a PSTN signalling message. It uses the INFO method to transmit such signalling to the egress gateway. The INFO method does not have to modify any parameter of the SIP session to enable the exchange of PSTN messages between gateways.

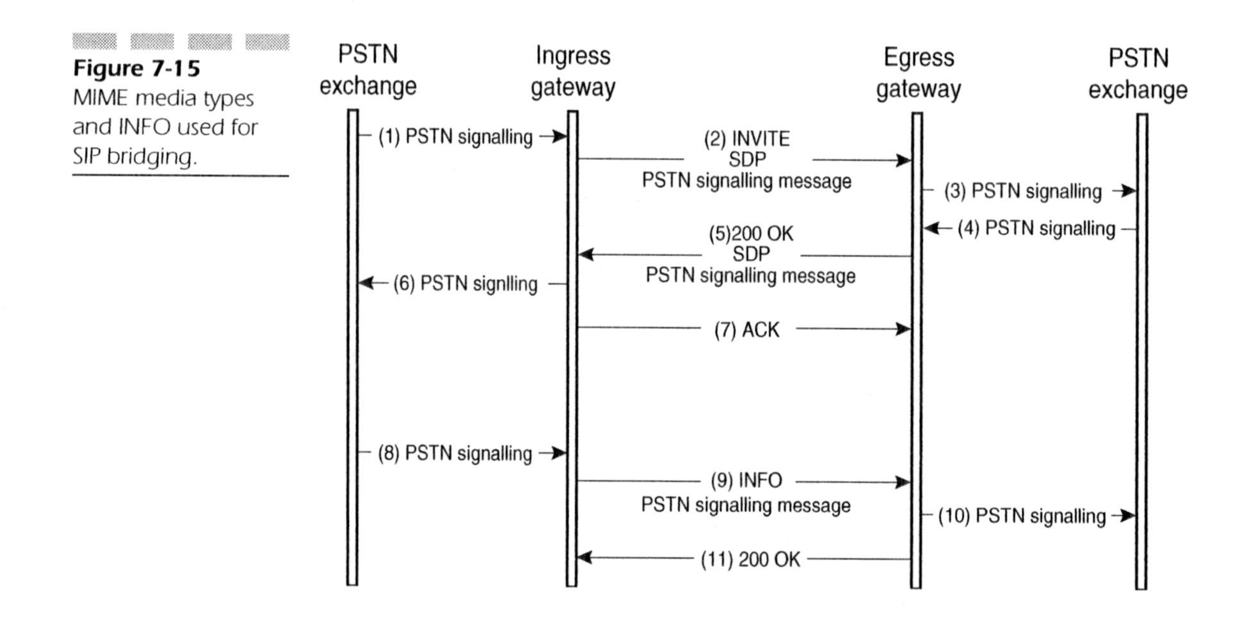

Figure 7-15 MIME media types and INFO used for SIP bridging.

The PINT Service Protocol

Another type of interworking with the PSTN is defined by PINT. The IETF *PSTN and Internet Interworking* (PINT) working group studies how Internet devices request PSTN telephony services. For instance, a user can click on a link in a Web page to request that someone from the support department call him or her on the PSTN phone. Or a user can request that a fax be sent to a certain machine on the PSTN providing the data, possibly though a pointer, to be faxed.

The PINT working group developed the PINT protocol [RFC 2848] in order to provide this kind of service. The protocol consists of SIP and SDP with extensions. Among other extensions, PINT uses the SUBSCRIBE/NOTIFY methods to receive status reports about the services invoked.

PINT's goal is to use SIP in a limited way to establish only those sessions that do not fall into the category of Internet sessions. PINT defines the SDP format to describe a fax session, for instance. Once this format is defined, SIP is used as usual to establish a session. The only difference is that the session in question happens to be a fax over the PSTN rather than *Real-time Transport Protocol* (RTP) packets over the Internet. Once again we can see the strength of separating session description from session establishment. The same protocol, SIP, can be used to establish any kind of session. The only requirement is that a session description format that can be carried in a SIP body must be present to describe it.

Figure 7-16 shows how Bob uses SIP to fax Laura. Bob does not have a fax machine in his office, but the information she needs is on ftp://ftp.company.com/Bob/document.txt. Using SIP, he sends an INVITE to a PINT gateway that specifies the information to be sent (through a *Universal Resource Identifier* (URI)) and bears Laura's fax machine number (1-212-555-5555). The PINT gateway interfaces the Internet and the PSTN. It downloads the information from the URI provided and sends it to Laura's fax machine, while it sends notifications to Bob about the status of the fax.

The SDP syntax to describe this service is pretty simple. A new address type (*c* line) has been defined as *Telephone Network* (TN). It contains a PSTN phone number. The *m* lines describe the media stream (voice, fax, or pager).

```
c= TN RFC2543 +1-212-555-5555

m= text 1  fax plain

a=fmtp:plain  uri:ftp://ftp.company.com/~Bob/document.txt
```

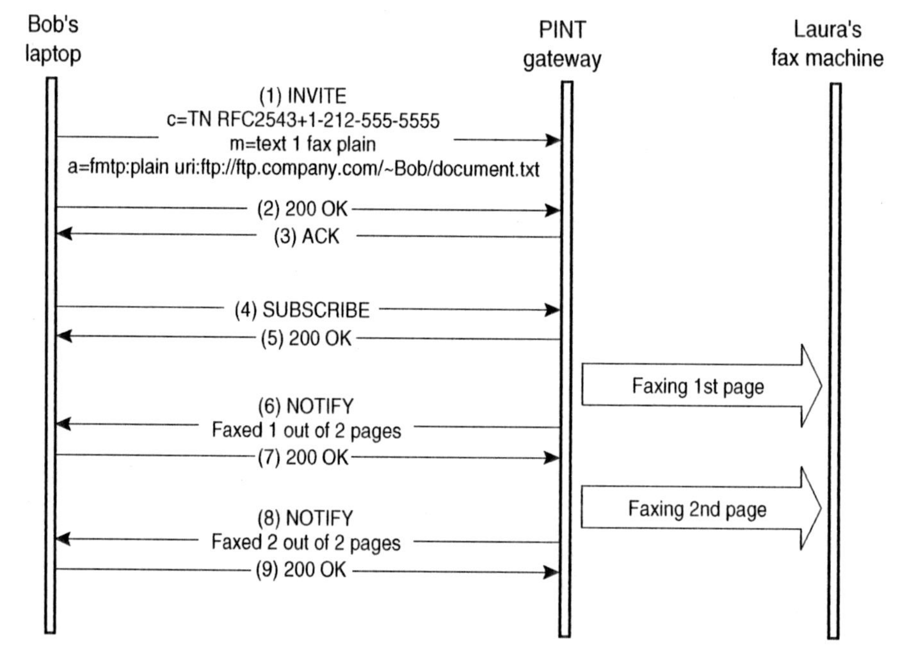

Figure 7-16 Fax session over the PSTN established by SIP.

SIP for Conferencing

We know that many services that have been implemented using SIP consist of point-to-point sessions. When Bob establishes a session with Laura, they engage in a two-party session where signalling and media is exchanged directly between them and no one else. However, two-party sessions are not what SIP does best. In fact, we saw that SIP and SDP were originally designed for conferences on the MBONE where members of a multicast group receive media on a multicast address.

Multiparty conferences form the basis of many services. Probably the easiest example is a videoconference with three or more members. In this case, exchanging media between persons at the conference is synonymous with being conferenced.

Nothing mandates that conference members must all be humans, though. Conferencing is used in many situations where only two humans are involved. For instance, Bob and Laura can have a business meeting over their SIP phones and record their conversation so Laura's assistant can later prepare the meeting minutes. In this scenario, Bob and Laura are in fact on a multiparty conference because the recording machine is a third member of the session.

Conferencing mechanisms [draft-rosenberg-sip-conferencing-models] are most powerful when seen as service enablers rather than just a means for a large group of people to communicate. The introduction of non-human conference members can provide many different services, including background music and image manipulation.

In the following section, we will describe several conferencing models supported by SIP and how they are implemented.

Multicast Conferences

The first thing to notice is that although members of a multicast group receive media on a multicast address, SIP signalling is point to point between session participants.

For instance, Bob can send an INVITE to Laura with the description of a multicast session. Bob and Laura use traditional unicast to send and receive SIP messages, but use multicast to send and receive media. Therefore, a point-to-point protocol like SIP can be used to signal multicast conferences.

Multicast conferences are usually prearranged events and thus, SIP is not used to create them or to terminate them. Accordingly, a SIP INVITE for a multicast conference is used only to INVITE users and not to establish the conference session. SIP BYEs are also not used when a user exits the conference. Note that each participant typically has a SIP signalling relationship with just one other participant, not with all the participants in the conference.

Multicast conferences work well for prearranged events with a large number of participants. However, for smaller events or for ad-hoc conferences, obtaining multicast addresses to deliver the media is not worthwhile. Hence, SIP supports other conferencing models that don't scale as well but provide other advantages.

End User Mixing Model

It is common for a two-party session to add participants as time goes by. The most straightforward way to conference in a new participant is for a current member of the session to INVITE him or her. The member who issues the

INVITE will receive media from both the previous member and the new member. He or she will then have to mix the media and send the result to both of them.

This mechanism, depicted in Figure 7-17, works when the user performing the mixing is also the last person to leave the conference. But as the number of participants grows, the processing power required to mix the media from all participants will quickly exceed the user's equipment capabilities. So the end user mixing model is great for small ad-hoc conferences, but it cannot successfully be extended to larger sessions. Other conference models supported by SIP, notably those using a *Multipoint Control Unit* (MCU), are preferable.

Multipoint Control Unit (MCU)

MCUs are used for both prearranged and ad-hoc conferences. In a prearranged conference, users INVITE the MCU and the MCU mixes the media and sends it to all of them, establishing a signalling relationship between the MCU and each participant.

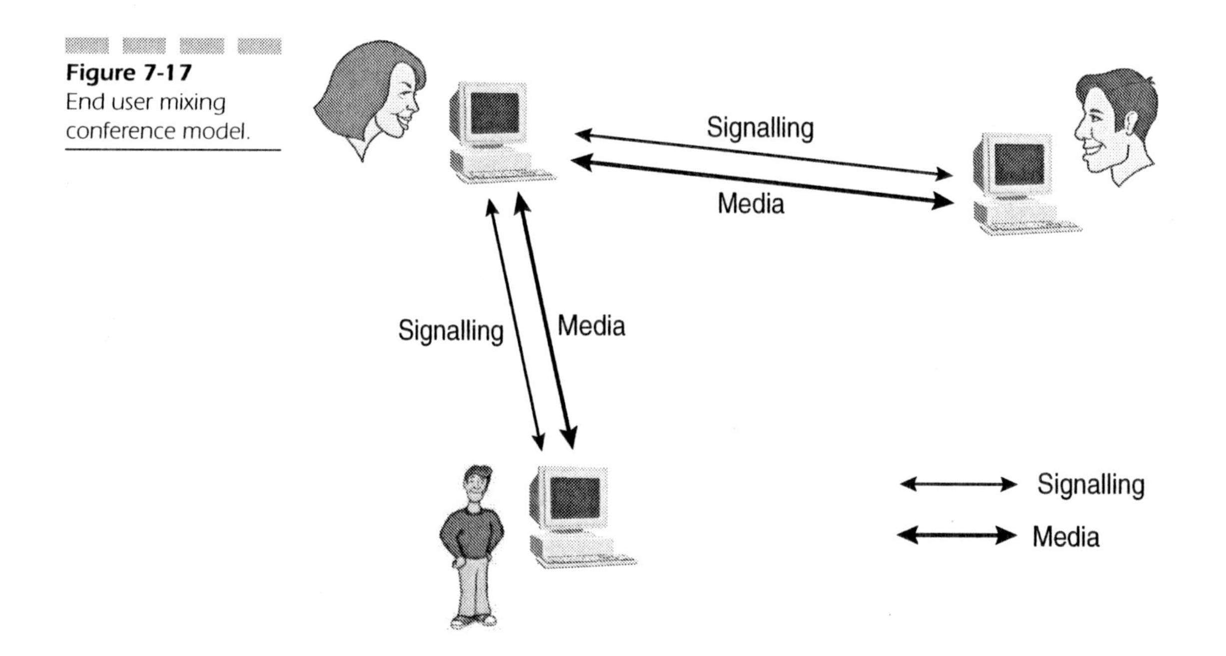

Figure 7-17 End user mixing conference model.

Alternatively, instead of having users invite the MCU, the MCU can INVITE participants to join a conference—a tactic particularly useful for prearranged events, where the MCU can INVITE all the users at once. In either case, the MCU handles both signalling and media for the conference (Figure 7-18).

An MCU-based conference scales better than an end user mixing conference. MCUs are prepared to handle media mixing and boast high processing capacity. However, an end user mixing conference can be converted into an MCU conference by means of the REFER method. In order to migrate from one model to the other, users are instructed to INVITE the MCU.

Decentralized Multipoint Conference

When media is sent in a decentralized fashion, it constitutes a special case of using a central signalling point such as an MCU. The conference unit handles signalling only when users send media using unicast or multicast (Figure 7-19). When multicast is used, the conference unit returns an SDP

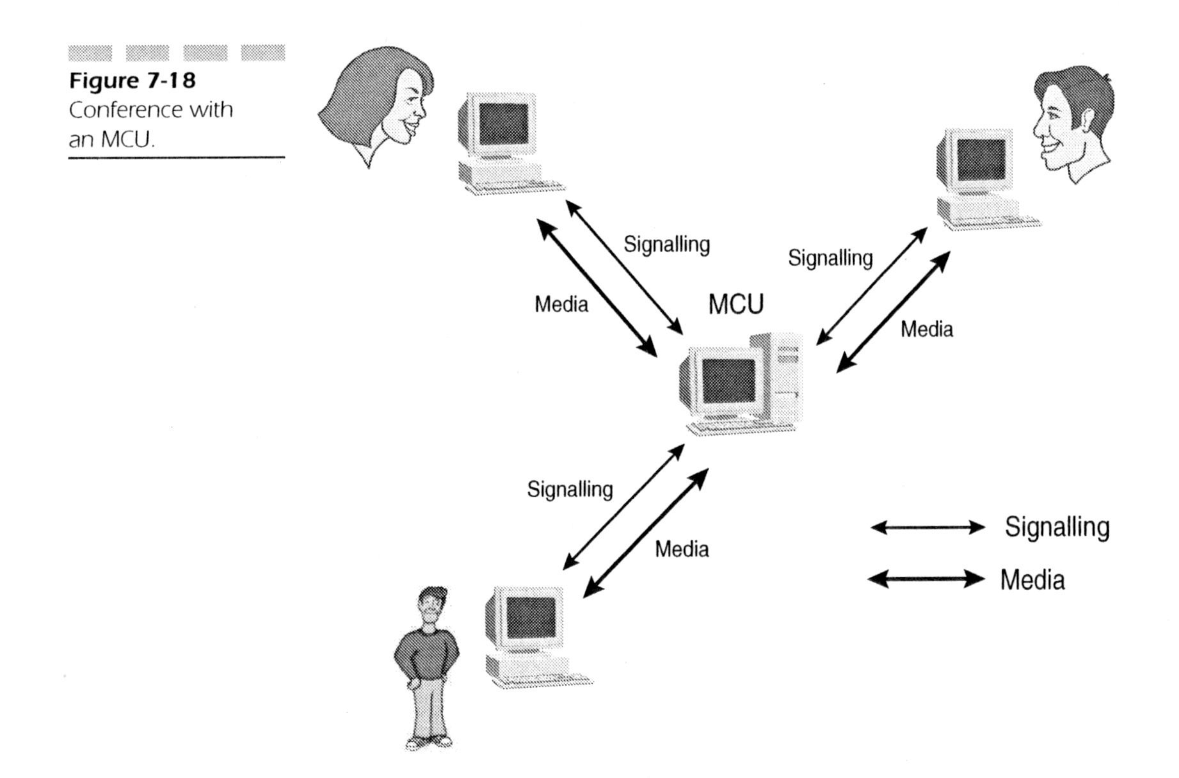

Figure 7-18 Conference with an MCU.

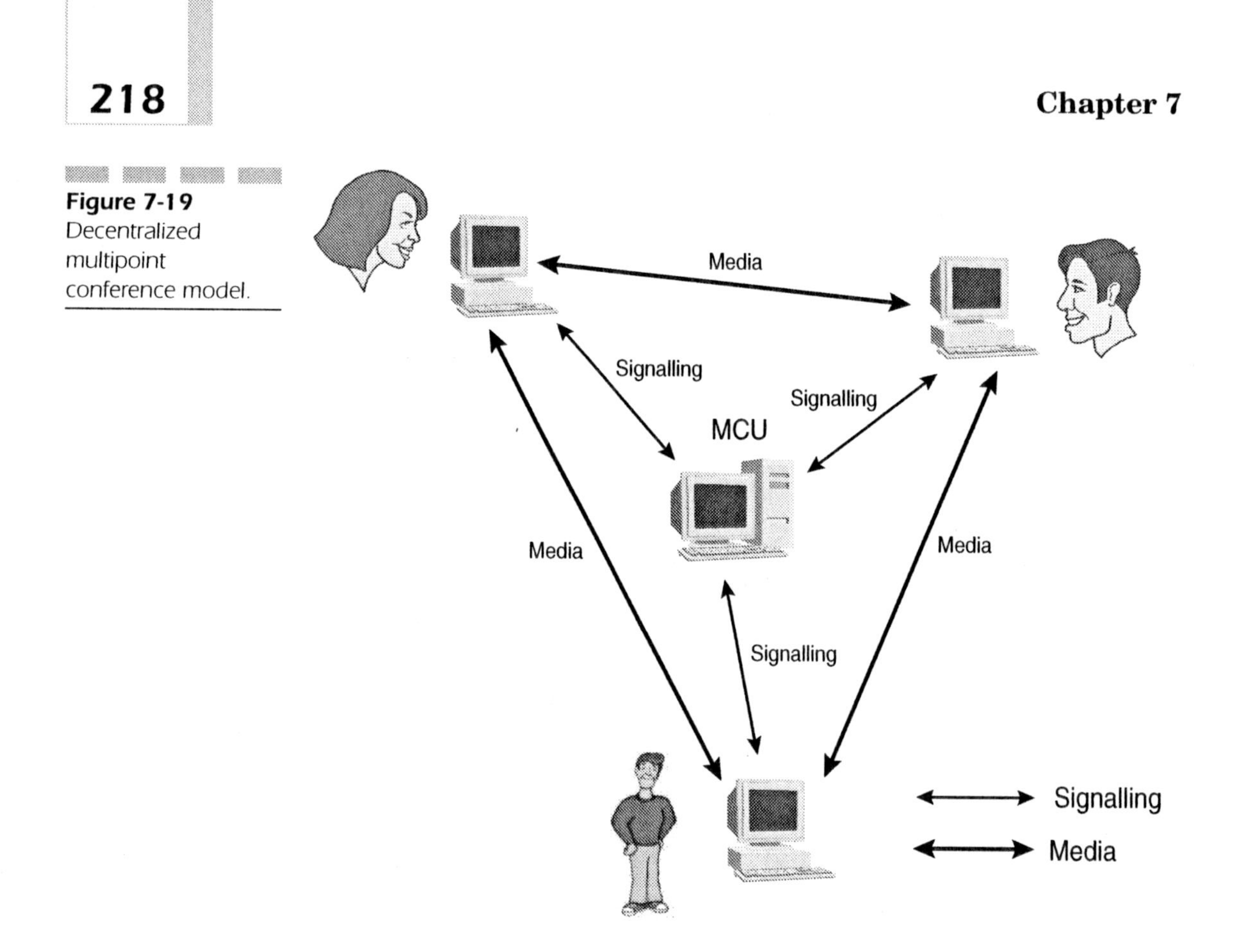

Figure 7-19 Decentralized multipoint conference model.

description to each participant with the multicast address that will be used. In the case of unicast, the conference unit adds a new *m* line though a re-INVITE every time a new participant joins the conference to trigger all current users to send media to the new participant as well.

This model simplifies the building of conference units because it obviates the need to handle media. In the case of multicast media, this model scales very well and a conference can handle many signalling relationships. When unicast media is used, though, the end systems have to send the same media in parallel to several different locations. End systems with low processing power might experience difficulty as the number of participants grows and choose to migrate to an MCU that handles media as well.

Control of Networked Appliances

Currently, the number of home devices with a network interface is relatively small: a desktop computer, perhaps a laptop, a printer, and so on. This situation is expected to change dramatically in the next few years. When everyday appliances such as the refrigerator or the alarm clock include a network interface, they become networked appliances. The main advantage of networked appliances is their capability to interact with one another and with any other networked device. Thus, when the refrigerator is running out of mayonnaise, it can order a fresh supply through the supermarket's Web page. The bedside clock can get information about traffic conditions and reset the alarm accordingly. If Bob forgets to turn off the porch light when he leaves home in the morning, he can always turn it off from his office.

A number of protocols are already designed to control networked appliances within a home. However, no solution is available for the inter-domain communication of network appliances [draft-tsang-appliances-reqs]. SIP, suitably extended, is a candidate to perform this role. Figure 7-20 shows how various SIP extensions work together to control the appliances inside a house. Say that a technician is scheduled to fix Bob's washing machine in the morning when Bob is at work. Bob simply SUBSCRIBEs to his doorbell. When the technician rings the bell, he automatically causes a NOTIFY to go to Bob. Bob then establishes a video session with the camera on his door, and checks that the person ringing is the technician. He also establishes an audio session with the audio system on his door and informs the technician that no one is home. Bob explains the trouble he's been having with the washing machine and gives the go-ahead for repairs. At this point, Bob sends a DO to his door in order to unlock it. Now the technician can do his work while Bob observes him through cameras installed in his place. When the technician finishes the job and leaves the apartment, Bob sends another DO to relock his door.

In this scenario, one of the main services SIP provides is security. Every command sent by Bob has to be strongly authenticated before it is performed. Bob definitely does not want anyone else to use the cameras in his place or to unlock his door. The SIP security framework provides mechanisms to control networked appliances while ensuring the privacy of the user.

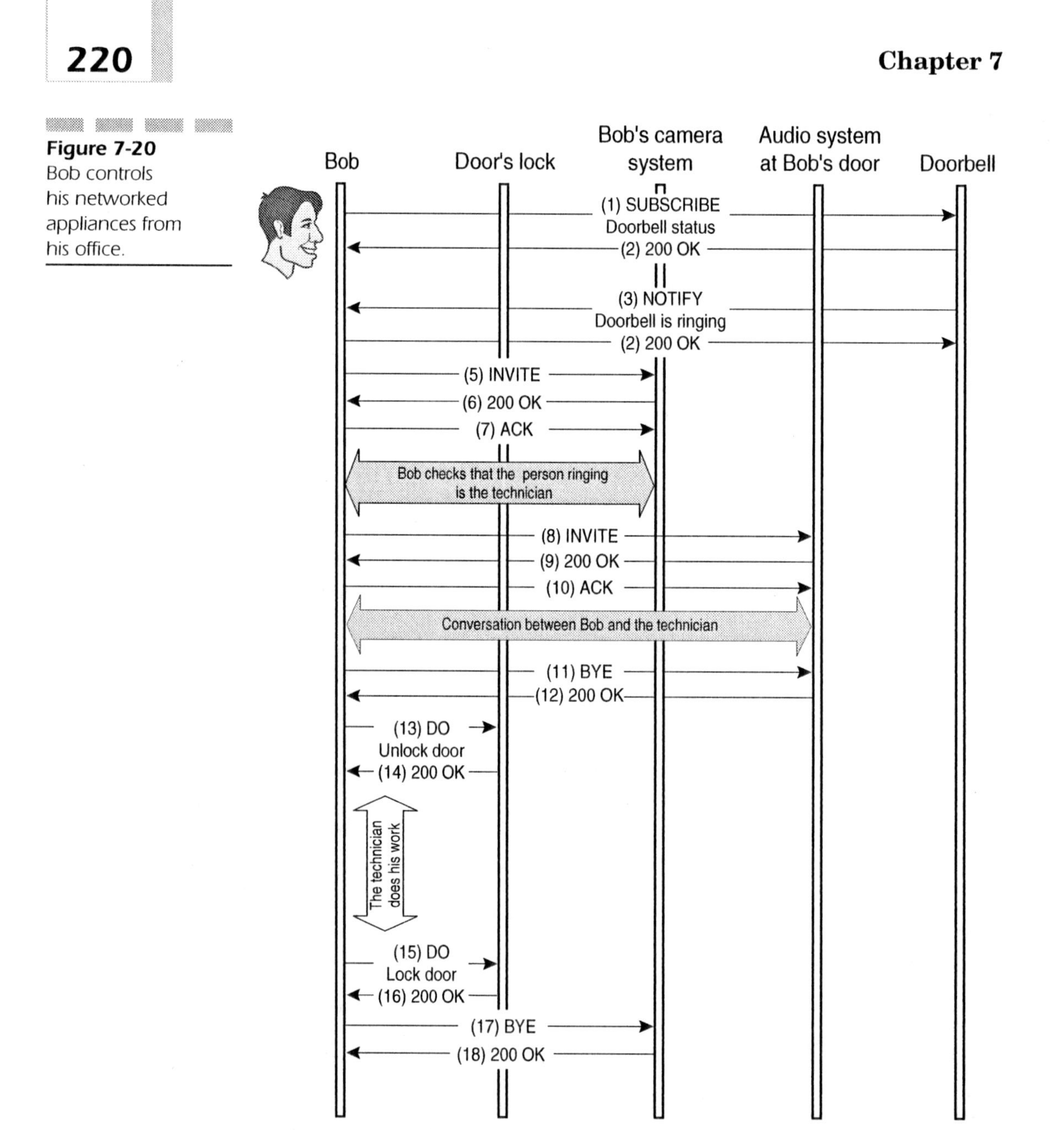

Figure 7-20 Bob controls his networked appliances from his office.

This chapter has provided some examples of architectures that use SIP as a signalling protocol. Now it is clear for the reader how to combine different SIP extensions, and even different protocols in order to provide real-life services to real users.

APPENDIX

Finding Futher Information on SIP

This appendix is for those of you that have finished this book and would like to read even more about SIP. The best way to find information about SIP or any other protocol developed by the IETF is to surf the Internet. However, if you are not familiar with IETF protocols, you might spend more time than you would like, looking for a particular piece of information on the Web. Finding a good Web page on a protocol is sometimes difficult. A regular search engine will return too many hits if you just introduce the word "SIP," for instance. This appendix contains the most useful Web pages related to SIP. These Web pages will help readers find the information they want more rapidly. We also give information about the most interesting SIP-related mailing lists. As we mentioned before, mailing lists are one of the most important tools in the IETF protocols development.

At the end of this appendix, we provide the readers with an example of an RFC. This will help them understand what an IETF specification looks like.

IETF Web site

The best place to find any specification of an Internet protocol is, of course, the IETF Web site (see Figure A-1).

The IETF Web page (http://www.ietf.org) contains information about the organization itself and about past and future meetings. However, we are more interested in finding technical specifications. We have given a number of references to Internet drafts and RFCs throughout the book. Finding any of those documents on this Web site is pretty easy. Under the link "Internet-Drafts," we will be able to look for a draft by the name of its author or title. There is also an index with all the Internet drafts classified by the working group they belong to.

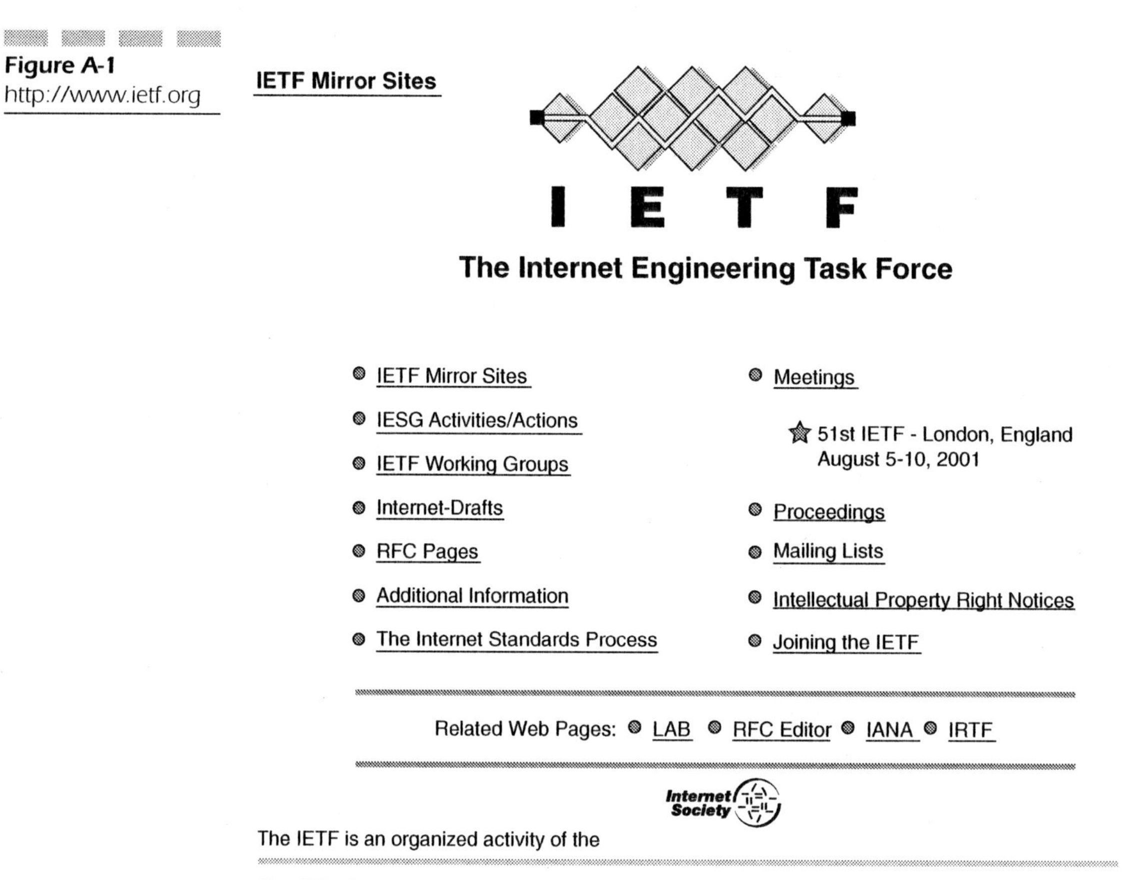

Figure A-1 http://www.ietf.org

If we are after a particular RFC rather than an Internet draft, we will find what we need under the link "RFC Pages." There we can look for a certain RFC number or read through the RFC index, which contains all the RFCs published by the IETF.

Any reader who is used to looking for information in other standardization bodies' Web pages will soon notice that the IETF does not require any kind of password or membership information in order to download any specification—they are available for free to any individual. Finding documents on the IETF Web page is much simpler than in other organizations, where it is difficult to keep track of the different versions of a certain document.

Readers might also find it useful to surf through the "IETF Working Groups" link. From there it is possible to access the Web pages of the different working groups. The reader will especially find the charters of SIP, SIMPLE, and MMUSIC very interesting. Another working group that might be of interest is IPTEL.

The Web page of a working group also contains instructions about how to join the mailing list of that specific working group. Reading the archives of any mailing list will provide the reader with several examples of the technical discussions that are carried out within the IETF community.

Note however that the mailing list of a working group is intended to be used by the engineers that are developing the protocol. The mailing list is not the proper place for people who are not very familiar with SIP to post basic questions about the protocol. There exists another mailing list for general information and discussions of existing implementations called "SIP implementors." Information about how to get subscribed to this mailing list can be also found in the SIP working group Web page (http://www.ietf.org/html.charters/sip-charter.html). In the next section, we will see where to find a useful set of *Frequently Asked Questions* (FAQ).

Henning Schulzrinne's SIP Web page

We have seen that the IETF Web site is the best place to download protocol specifications and to get subscribed to mailing lists. However, if you are after general SIP information, Henning Schulzrinne's Web page (Figure A-2) is the one you are looking for (http://www.cs.columbia.edu/sip).

Schulzrinne, one of the co-author's of SIP, maintains this Web page and it is unarguably the best page to follow SIP development. The "news" section contains information about SIP-related important events such as the release of a new RFC, or links to the organizers of the next SIP interoperability event. Schulzrinne's Web page also contains links to different organizations making use of SIP and to different presentations given in different telecommunications conferences. It is worthwhile to spend some

Figure A-2
http://www.cs.columbia.edu/sip

Session Initiation Protocol (SIP)

SIP, the Session Initiation Protocol, is a signaling protocol for Internet conferencing, telepony, presence, events notification and instant messaging. SIP was developed within the IETF MMUSIC (Multiparty Multimedia Session Control) working group, with work proceeding since September 1999 in the IETF SIP working group.

A number of standardization organizations and groups are using or considering SIP:

- IETF PINT working group
- 3GPP for third-generation wireless networks
- Softswitch Consortium
- IMTC and ETSI Tiphon are working onSIP-H.323 interworking
- PacketCable DCS (distributed call signaling) specification
- 3GPP (for third-generation wireless)
- SpeechLinks, for moving between speech-enabled sites

News

- May 29, 2001: RFC2543bis (-03) draft
- May 4, 2001: Information about the 8th SIP Interoperability Test Event is now available
- April 14, 2001: search feature added.
- April 11, 2001: The SIP interoperability test event has a new logo, courtesy of Ubiquity.

- April 10, 2001: RFC 3087 (*Control of Service Context using SIP Request-URI*) published
- Feb. 1, 2001: RFC 3050 (*Common Gateway Interface for SIP*) published
- Nov. 30, 2000: Caller preferences draft in WG last call until December 24, 2000
- Nov. 29, 2000: Guidelines for Authors of SIP Extensions draft in WG last call until December 24, 2000
- November 24, 2000: RFC2543bis (-02) draft
- Nov. 17, 2000: CPL in IESG last call.
- RFC 2976 (*The SIP INFO Method*) published
- The sixth SIP interoperability test event took place December 5-8, 2000 at Sylantro and Sun in Silicon Valley, California.
- June 20, 2000: The SIP Forum was founded. "SIP Forun is a non profit association whose mission is to promote awareness and provide information about the benefits and capabilities that are enabled by SIP."
- The fifth SIP interoperability test event took place August 8-10, 2000 at pulver.com in Melville, Long Island.
- June 15, 2000: RFC 2848, *The PINT Service Protocol: Extensions to SIP and SDP for IP Access to Telephone Call Services*, published.
- Added SIP internship and job listing.
- The fourth SIP interoperability test event took place April 17-19, 2000 in Rolling Meadows (near Chicago), Illinois, hosted by 3Com.
- February 28, 2000): Draft *The SIP INFO Method* is in IETF last call for Proposed Standard.
- September 1999: A new IETF working group on SIP has been created.
- The third SIP interoperability test event took place December 6th throught 8th, 1999 in Richardson, Texa, hosted by Ericsson.
- The second SIP interoperability test event took place August 5th and 6th, 1999 at pulver.com (Melville, NY).
- The first SIP interoperability test event (known as "bake off") took place April 8th and 9th, 1999 at Columbia University, New York.
- SIP is a Proposed Standard (Feb. 2, 1999) published as RFC 2543 (March 17, 1999).
- New list of public SIP servers.

[Overview] [Where is SIP being discussed?] [Search] [What SIP extensions are being planned?] [Drafts] [SIP grammar] [Implementations] [SIP service providers] [mailing list] [Public SIP Servers] [Papers] [Talks] [Related Drafts and Documents] [Draft History] [Frequently Asked Questions (FAQ)] [Port Assignments and DNS] [Compact headers] [SIP-related events] [SIP interoperability testing events] [Press Coverage] [Emergency Calling] [Non-Internet-related SIP sightings] [Status and Schedule] [Internships and Jobs] [Other SIP sites]

HITOMETER
316045

Last updated Friday, June 15, 2001 21:30:07 by Hennig Schulzrinne

time surfing through this Web page to get familiar with it. All the information that one needs about SIP can be found here.

Of special interest is the FAQ section. Any newcomer to SIP who has any question and is thinking of posting a message to the general information mailing list should first check the FAQ section. The most common questions with their answers are gathered here. Checking the FAQs first before posting questions to a mailing list does not just save some bandwidth. It also saves time to the engineers that are subscribed to that mailing list in order to help people to get familiar with the protocol. They very much appreciate not having to answer the same question several times.

Dean Willis' Web Pages

A link to Dean Willis' Web page can be found in the SIP working group official Web page at the IETF. This additional SIP Web page (http://www. softarmor.com/sipwg) is maintained by Dean Willis, co-chairman of the SIP working group. This Web page (Figure A-3) deals with the administrations of the SIP working group. It has information about the small design teams, groups of people that work together to resolve a specific issue, and about the final call calendar.

The final call calendar is very important because by looking at it, we can know when an Internet draft will become an RFC. This is the best place to check the maturity of different extensions. When a draft enters in its final call, the working group has a last chance to comment on it before the working group sends the draft to the IESG. Once the draft is submitted to the IESG, the area directors will decide if it can become an RFC or if the working group needs to continue working on it longer.

Dean Willis also maintains another Web page (Figure A-4) at http://www. softarmor.com/sipping. This Web page contains work leading to specify frameworks and requirements for new SIP-based applications. There is also a last call calendar with the status of all the related Internet drafts.

Figure A-3
http://softarmor.com/sipwg

Session Initiation Protocol (SIP) Working Group Supplemental Home Page

[Drafts] [Morgue] [Last Call Info] [Meetings] [Design teams] [Issues] Process] [References] [SIPPING]

Important Notes

14-Jun-01: the Last Call Calendar has been updated.

1-Jun-01: This site has been heavily modified.

1-Jun-01: See the SIPPING Site here.

General Notes

Volunteer to conduct nits and last call reviews by contacting Rakesh Shah (rshah@dynamicsoft.com).

Menu

- Drafts – Text and status information.
- Morgue – Archive of expired drafts.
- Last Call Info – Schedule, reviewers and reviewers references.
- Meetings – IETF, Interim Meetings, etc.
- Design Teams – Teams working on specific issues.
- Issues – Stuff we need to take care of.
- Processes – Ok, everybody has to have a few rules.
- References – Some useful information that we don't have.
- SIPPING – SIPPING Working Group Supplemental Home Page.

Brought to you as a personal effort (that's no corporate sponsorship implied, ok?) of Dean Willis, who can be reached at or dean.willis@softarmor.com and serveral other unlikely places. This is a personal server. Please don't thrash it. Have a nice day!

Updated June 06, 2001 14:08 UST

The SIP forum

Although the SIP forum (Figure A-5) is not a technical organization, it is worthwhile mentioning. The purpose of the SIP forum (http://www.sipforum.org) is to spread information related to SIP. This is usually done through papers, conferences, and mailing lists. It is important to note that the SIP forum does not produce any technical specification, it just spreads information about SIP.

Figure A-4
http://softarmor.com/sipping

Session Initiation Protocol INvestiGation (SIPPING) Working Group
Supplemental Home Page

[Drafts] [Morgue] [Last Call Info] [Meetings] [Design Teams] [Issues] [Process] [References] [SIP]

Important Notes

14-Jun-01: The Last Call Calendar has been updated.

1-Jun-01: This site is all new.

1-Jun-01: See the SIP Site here

General Notes

Volunteer to conduct nits and last call reviews by contacting Rakesh Shah (rshah@dynamicsoft.com).

Menu

- Draft charter for SIPPING Working Group.
- Drafts -- Text and status information.
- Morgue -- Archive of expired drafts.
- Last Call Info -- Schedule, reviewers and reviewers references.
- Meetings -- IETF, Interim Meetings, etc.
- Design Teams -- Teams working on specific issues.
- Issues -- Stuff we need to take care of.
- Processes -- Ok, everybody has to have a few rules.
- References -- Some useful information that we don't have.
- SIP WG -- SIP Working Group Supplemental Home Page.

Brought to you as a personal effort (that's no corporate sponsorship implied, ok?) of Dean Willis, who can be reached at or dean.willis@softarmor.com and several other unlikely places. This is a personal server. Please don't thrash it. Have a nice day!

Updated June 06, 2001 14:08 UST

RFC example

This section consists of an example of an RFC. This is intended for readers that do not feel comfortable surfing the Internet looking for information but would like to know what an RFC looks like. Since this section is intended only to be an example of an IETF specification, we have chosen the shortest (in number of pages) RFC related to SIP that has been released so far: the SIP INFO method [RFC 2976].

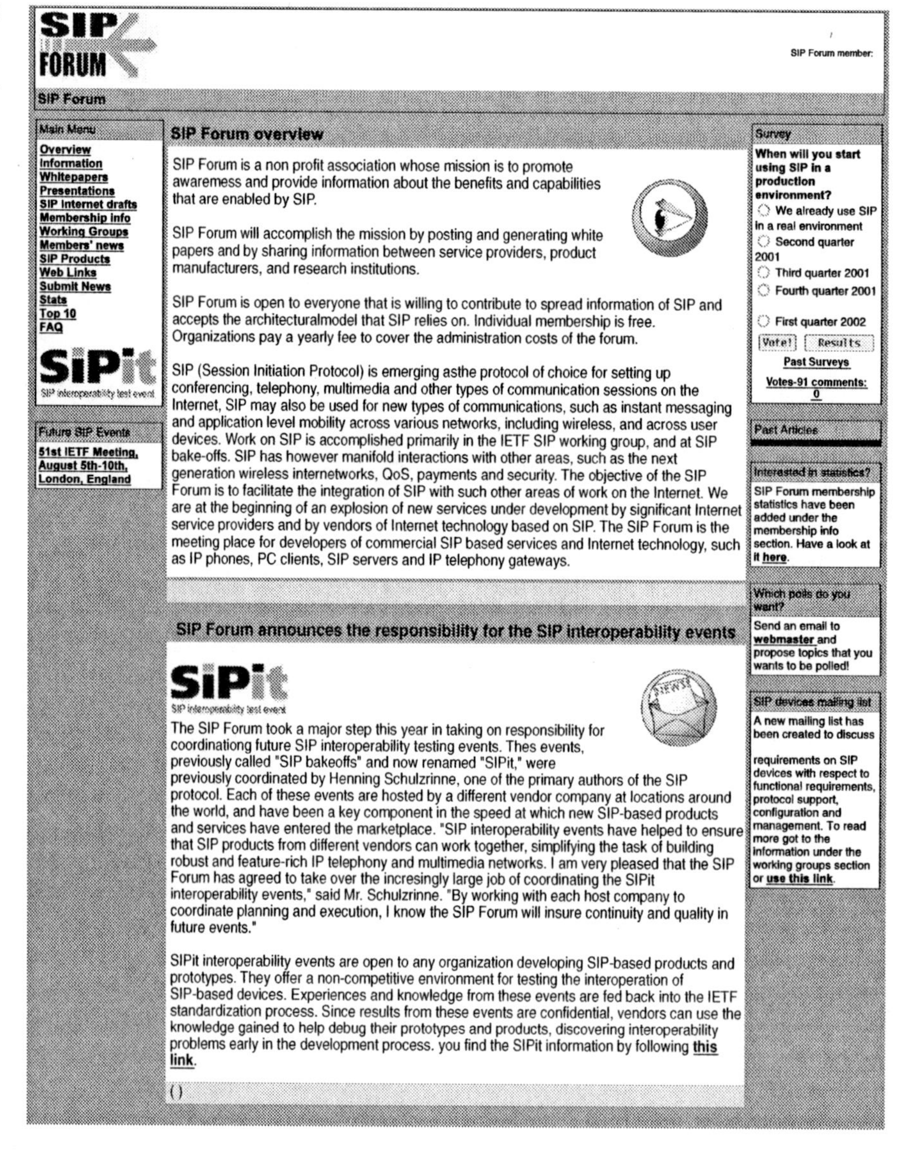

Figure A-5
http://sipforum.org

The readers might find it interesting to read the full copyright statement in the last page of this RFC. Each RFC contains such a statement at the end. Another interesting aspect of each RFC is that they all are written in textual format. Therefore, if they include any figures they are always drawn using textual characters.

This RFC can be downloaded from http://www.ietf.org/rfc/rfc2976.txt

RFC

Network Working Group
Request for Comments: 2976
Category: Standards Track

S. Donovan
dynamicsoft
October 2000

The SIP INFO Method

Status of this Memo

This document specifies an Internet standards track protocol for the Internet community, and requests discussion and suggestions for improvements. Please refer to the current edition of the "Internet Official Protocol Standards" (STD 1) for the standardization state and status of this protocol. Distribution of this memo is unlimited.

Copyright Notice

Abstract

This document proposes an extension to the Session Initiation Protocol (SIP). This extension adds the INFO method to the SIP protocol. The intent of the INFO method is to allow for the carrying of session related control information that is generated during a session. One example of such session control information is ISUP and ISDN signaling messages used to control telephony call services.

This and other example uses of the INFO method may be standardized in the future.

Table of Contents

1. Introduction

The SIP protocol described in [1] defines session control messages used during the setup and tear down stages of a SIP controlled session.

In addition, the SIP re-INVITE can be used during a session to change the characteristics of the session. This is generally to change the properties of media flows related to the session or to update the SIP session timer.

However, there is no general-purpose mechanism to carry session control information along the SIP signaling path during the session.

The purpose of the INFO message is to carry application level information along the SIP signaling path.

The INFO method is not used to change the state of SIP calls, or the parameters of the sessions SIP initiates. It merely sends optional application layer information, generally related to the session.
It is necessary that the mid-session signaling information traverse the post session setup SIP signaling path. This is the path taken by SIP re-INVITEs, BYEs and other SIP requests that are tied to an individual session. This allows SIP proxy servers to receive, and potentially act on, the mid-session signaling information.

This document proposes an extension to SIP by defining the new INFO method. The INFO method would be used for the carrying of mid-call signaling information along the session signaling path.

1.1 Example Uses

The following are a few of the potential uses of the INFO message:

- Carrying mid-call PSTN signaling messages between PSTN gateways,

- Carrying DTMF digits generated during a SIP session.

- Carrying wireless signal strength information in support of wireless mobility applications.

- Carrying account balance information.

- Carrying images or other non streaming information between the participants of a session.

These are just potential uses; this document does not specify such uses nor does it necessarily recommend them.

It can also be envisioned that there will be other telephony and non-telephony uses of the INFO method.

2. INFO Method

The INFO method is used for communicating mid-session signaling information along the signaling path for the call.

The INFO method is not used to change the state of SIP calls, nor does it change the state of sessions initiated by SIP. Rather, it provides additional optional information which can further enhance the application using SIP.

The signaling path for the INFO method is the signaling path established as a result of the call setup. This can be either direct signaling between the calling and called user agents or a signaling path involving SIP proxy servers that were involved in the call setup and added themselves to the Record-Route header on the initial INVITE message.

The mid-session information can be communicated in either an INFO message header or as part of a message body. The definition of the message body and/or message headers used to carry the mid-session information is outside the scope of this document.

There are no specific semantics associated with INFO. The semantics are derived from the body or new headers defined for usage in INFO.

2.1 Header Field Support for INFO Method

Tables 1 and 2 add a column to tables 4 and 5 in the [1]. Refer to Section 6 of [1] for a description of the content of the tables. Note that the rules defined in the enc. and e-e columns in tables 4 and 5 in [1] also apply to use of the headers in the INFO request and responses to the INFO request.

2.2 Responses to the INFO Request Method

If a server receives an INFO request it MUST send a final response.

A 200 OK response MUST be sent by a UAS for an INFO request with no message body if the INFO request was successfully received for an existing call. Beyond that, no additional operations are required.

Header	Where	INFO
Accept	R	o
Accept-Encoding	R	o
Accept-Language	R	o
Allow	200	-
Allow	405	o
Authorization	R	o
Call-ID	gc	m
Contact	R	o
Contact	1xx	-
Contact	2xx	-
Contact	3xx	-
Contact	485	-
Content-Encoding	e	o
Content-Length	e	o
Content-Type	e	*
CSeq	gc	m
Date	g	o
Encryption	g	o
Expires	g	o
From	gc	m
Hide	R	o
Max-Forwards	R	o
Organization	g	o

Table 1 Summary of header fields, A-0

Handling of INFO messages that contain message bodies is outside the scope of this document. The documents defining the message bodies will also need to define the SIP protocol rules associated with those message bodies.

A 481 Call Leg/Transaction Does Not Exist message MUST be sent by a UAS if the INFO request does not match any existing call leg.

If a server receives an INFO request with a body it understands, but it has no knowledge of INFO associated processing rules for the body, the body MAY be rendered and displayed to the user. The INFO is responded to with a 200 OK.

If the INFO request contains a body that the server does not understand then, in the absence of INFO associated processing rules for the body, the server MUST respond with a 415 Unsupported Media Type message.

Header	Where	INFO
------	-----	----
Priority	R	o
Proxy-Authenticate	407	o
Proxy-Authorization	R	o
Proxy-Require	R	o
Require	R	o
Retry-After	R	-
Retry-After	404,480,486	o
Retry-After	503	o
Retry-After	600,603	o
Response-Key	R	o
Record-Route	R	o
Record-Route	2xx	o
Route	R	o
Server	r	o
Subject	R	o
Timestamp	g	o
To	gc(1)	m
Unsupported	420	o
User-Agent	g	o
Via	gc(2)	m
Warning	r	o
WWW-Authenticate401	o	

Table 2 Summary of header fields, P-Z

Bodies which imply a change in the SIP call state or the sessions initiated by SIP MUST NOT be sent in an INFO message.

Other request failure (4xx), Server Failure (5xx) and Global Failure (6xx) responses MAY be sent for the INFO Request.

2.3 Message Body Inclusion

The INFO request MAY contain a message body.

2.4 Behavior of SIP User Agents

Unless stated otherwise, the protocol rules for the INFO request governing the usage of tags, Route and Record-Route, retransmission and reliability, CSeq incrementing and message formatting follow those in [1] as defined for the BYE request.

An INFO request MAY be cancelled. A UAS receiving a CANCEL for an INFO request SHOULD respond to the INFO with a "487 Request Cancelled" response if a final response has not been sent to the INFO and then behave as if the request were never received.

However, the INFO message MUST NOT change the state of the SIP call, or the sessions initiated by SIP.

2.5 Behavior of SIP Proxy and Redirect Servers

2.5.1 Proxy Server

Unless stated otherwise, the protocol rules for the INFO request at a proxy are identical to those for a BYE request as specified in [1].

2.5.2 Forking Proxy Server

Unless stated otherwise, the protocol rules for the INFO request at a proxy are identical to those for a BYE request as specified in [1].

2.5.3 Redirection Server

Unless stated otherwise, the protocol rules for the INFO request at a proxy are identical to those for a BYE request as specified in [1].

3. INFO Message Bodies

The purpose of the INFO message is to carry mid-session information between SIP user agents. This information will generally be carried in message bodies, although it can be carried in headers in the INFO message.

The definition of the message bodies or any new headers created for the INFO method is outside the scope of this document. It is expected that separate documents will be created to address definition of these entities.

In addition, the INFO method does not define additional mechanisms for ensuring in-order delivery. While the CSeq header will be incremented upon the transmission of new INFO messages, this should not be used to determine the sequence of INFO information. This is due to the fact that there could be gaps in the INFO message CSeq count caused by a user agent sending re-INVITES or other SIP messages.

4. Guidelines for extensions making use of INFO

The following are considerations that should be taken into account when defining SIP extensions that make use of the INFO method.

- Consideration should be taken on the size of message bodies to be carried by INFO messages. The message bodies should be kept small due to the potential for the message to be carried over UDP and the potential for fragmentation of larger messages.

- There is potential that INFO messages could be forked by a SIP Proxy Server. The implications of this forking of the information in the INFO message need to be taken into account.

- The use of multi-part message bodies may be helpful when defining the message bodies to be carried by the INFO message.

- The extensions that use the INFO message MUST NOT rely on the INFO message to do anything that effects the SIP call state or the state of related sessions.

- The INFO extension defined in this document does not depend on the use of the Require or Proxy-Require headers. Extensions using the INFO message may need the use of these mechanisms. However, the use of Require and Proxy-Require should be avoided, if possible, in order to improve interoperability between SIP entities.

5. Security Considerations

If the contents of the message body are private then end-to-end encryption of the message body can be used to prevent unauthorized access to the content.

There are no other security issues specific to the INFO method. The security requirements specified in the SIP specification apply to the INFO method.

6. References

[1] Handley, M., Schulzrinne, H., Schooler, E. and J. Rosenberg, "SIP: Session Initiation Protocol", RFC 2543, March 1999.

7. Acknowledgements

The author would like to thank Matthew Cannon for his contributions to this document. In addition, the author would like to thank the members of the MMUSIC and SIP working groups, especially Jonathan Rosenberg, for comments and suggestions on how to improve the document.

8. Author's Address

Steve Donovan
dynamicsoft
5100 Tennyson Parkway, Suite 200
Plano, Texas 75024

E-mail: sdonovan@dynamicsoft.com

9. Full Copyright Statement

Acknowledgement

Funding for the RFC Editor function is currently provided by the Internet Society.

ACRONYMS

2G	Second generation of mobile systems
3G	Third generation of mobile systems
3GPP	Third Generation Partnership Project
3GPP2	Third Generation Partnership Project 2
AC	Alternating current
ANSI	American National Standards Institute
APEX	Application EXchange
ARIB	Association of Radio Industries and Businesses
ASCII	American Standard Code for Information Interchange
ASN-1	Abstract Syntax Notation 1
ATM	Asynchronous Transfer Mode
AVP	Audio/Video Profile
BCP	Best Current Practice
BOF	Birds of Feathers
BTS	Base Transceiver Station
CAS	Channel Associated Signalling
CCP	Connection Control Protocol
CMS	Call Management Server
CMSS	Call Management Server Signalling
CSCF	Call/Session Control Function
CSS	Common Channel Signalling
CWTS	China Wireless Telecommunications Standard
DC	Direct current
DHCP	Dynamic Host Configuration Protocol
DiffServ	Differentiated Services

DMP	Device Messaging Protocol
DSS-1	Digital Subscriber Line No. 1
DTMF	Dual Tone Multi-Frequency
EIA	Electronic Industries Alliance
ETSI	European Telecommunication Standard Institute
FDM	Frequency Division Multiplexing
GSM	Global System for Mobile communications
HSS	Home Subscriber Server
HTTP	Hypertext Transfer Protocol
IAB	Internet Architecture Board
ICB	Internet Cooperation Board
ICCB	Internet Configuration Control Board
I-CSCF	Interrogating-Call/Session Control Function
ID	Identification
I-D	Internet Draft
IETF	Internet Engineering Task Force
IMAP	Internet Message Access Protocol
IP	Internet Protocol
IPSec	Internet Protocol Security
IPTEL	IP Telephony
IMPP	Instant Messaging and Presence Protocol
IMT-2000	International Mobile Telecommunications 2000
IN	Intelligent Network
INAP	Intelligent Network Application Protocol
INRIA	Institut National de Recherche en Informatique et en Automatique
IRG	Internet Research Group

ISDN	Integrated Services Digital Network
ISOC	Internet Society
ISUP	ISDN User Part
ITU	International Telecommunication Union
ITU-T	International Telecommunication Union Telecommunication Standardization Sector
IVS	INRIA Videoconferencing System
Kbps	Kilobits per second
LAN	Local Area Network
LDAP	Lightweight Directory Access Protocol
MCU	Multipoint Control Unit
MIME	Multipurpose Internet Mail Extensions
MG	Media Gateway
MGC	Media Gateway Controller
MGCP	Media Gateway Control Protocol
MMCC	Multimedia Conference Control
MMUSIC	Multiparty Multimedia Session Control
MSC	Mobile Switching Center
MTA	Multimedia Terminal Adapter
NCP	Network Control Protocol
NCS	Network Call Signalling
PBX	Private Branch Exchange
PCM	Pulse Code Modulation
PDF	Portable Document Format
P-CSCF	Proxy-Call/Session Control Function
PHB	Per Hop Behavior
PINT	PSTN and Internet Interworking

PRIM Presence and Instant Messaging

PSTN Public Switched Telephone Network

PUA Presence User Agent

QoS Quality of Service

RFC Request for Comments

RFI Request for Information

RFP Request for Products

RSVP ReSerVation Protocol

RTCP Real-time Transport Control Protocol

RTP Real-time Transport Protocol

RTSP Real-time Streaming Protocol

SAP Session Announcement Protocol

SCIP Simple Conference Invitation Protocol

SCP Service Control Point

S-CSCF Serving-Call/Session Control Function

SCTP Stream Control Transmission Protocol

SDP Session Description Protocol

SDPng SDP next generation

SG Signalling Gateway

SIMPLE SIP for Instant Messaging and Presence Leveraging Extensions

SIP Session Initiation Protocol

SLP Service Location Protocol

S/MIME Secure/Multipurpose Internet Mail Extensions

SMTP Simple Mail Transport Protocol

SS6 Signalling System no. 6

SS7 Signalling System no. 7

SSP	Service Switching Point
STD	Standard
TCP	Transmission Control Protocol
TDM	Time Division Multiplexing
TIA	Telecommunications Industry Association
TLS	Transport Layer Security
TN	Telephone Network
TTA	Telecommunications Technology Association
TTC	Telecommunications Technology Committee
TUP	Telephone User Part
TV	Television
UA	User Agent
UAC	User Agent Client
UAS	User Agent Server
UDP	User Datagram Protocol
URI	Universal Resource Identifier
URL	Uniform Resource Locator
US	United States
VCR	Video Cassette Recorder
VoIP	Voice over IP

REFERENCES

[draft-ietf-bgmp-spec] D. Thaler, D. Estrin, D. Meyer, "Border Gateway Multicart Protocol (BGMP)," IETF. Work in progress.

[draft-ietf-impp-cpim] D. Crocker, A. Diacakis, F. Mazzoldi, C. Huitema, G. Klyne, M. Rose, J. Rosenberg, R. Sparks, H. Sugano. "A Common Profile for Instant Messaging (CPIM)," IETF. Work in progress.

[draft-ietf-mmusic-confarch] M. Handley, J. Crowcroft, C. Bormann, J. Ott. "The Internet Multimedia Conferencing Architecture," IETF. Work in progress.

[draft-ietf-mmusic-sdpng] D. Kutscher, J. Ott, C. Bormann. "Session Description and Capability Negotiation," IETF. Work in progress.

[draft-ietf-sip-100rel] J. Rosenberg, H. Schulzrinne. "Reliability of Provisional Responses in SIP," IETF. Work in progress.

[draft-ietf-sip-callerprefs] H. Schulzrinne, J. Rosenberg. "SIP Caller Preferences and Callee Capabilities," IETF. Work in progress.

[draft-ietf-sip-cc-transfer] R. Sparks. "SIP Call Control—Transfer," IETF. Work in progress.

[draft-ietf-sip-dhcp] G. Nair, H. Schulzrinne. "DHCP Option for SIP Servers," IETF. Work in progress.

[draft-ietf-sip-events] A. Roach, "Event Notification in SIP," IETF. Work in progress.

[draft-ietf-sip-guidelines] J. Rosenberg, H. Schulzrinne. "Guidelines for Authors of SIP Extensions," IETF. Work in progress.

[draft-ietf-sip-isup] G. Camarillo, A. Roach, J. Peterson, L. Ong. "ISUP to SIP Mapping," IETF. Work in progress.

[draft-ietf-sip-isup-mime] E. Zimmerer, J. Peterson, A. Vemuri, L. Ong, M. Watson, M. Zonoun. "MIME media types for ISUP and QSIG Objects," IETF. Work in progress.

[draft-ietf-sip-manyfolks-resource] W. Marshall, K. Ramakrishnan, E. Miller, G. Russell, B. Beser, M. Mannette, K. Steinbrenner, D. Oran, F. Andreasen, M. Ramalho, J. Pickens, P. Lalwaney, J. Fellows, D. Evans, K. Kelly, A. Roach, J. Rosenberg, D. Willis, S. Donovan, H. Schulzrinne. "Integration of Resource Management and SIP. SIP Extensions for Resource Management," IETF. Work in progress.

[draft-ietf-sip-rfc2543bis] M. Handley, H. Schulzrinne, E. Schooler, and J. Rosenberg. "SIP: Session Initiation Protocol," IETF. Work in progress.

[draft-ietf-sip-serverfeatures] J. Rosenberg, H. Schulzrinne. "The SIP Supported Header," IETF. Work in progress.

[draft-kempf-sip-findsrv] J. Kempf, J. Rosenberg. "Finding a SIP Server with SLP," IETF. Work in progress.

[draft-kutscher-mmusic-sdpng-req] D. Kutscher, J. Ott, C. Bormann. "Requirements for Session Description and Capability Negotiation," IETF. Work in progress.

[draft-moyer-sip-appliances-framework] S. Moyer, D. Marples, S. Tsang, J. Katz, P. Gurung, T. Cheng, A. Dutta, H. Schulzrinne, A. Roychowdhury. "Framework Draft for Networked Appliances Using the Session Initiation Protocol," IETF. Work in progress.

[draft-rosenberg-impp-im] J. Rosenberg, D. Willis, R. Sparks, B. Campbell, H. Schulzrinne, J. Lennox, C. Huitema, B. Aboba, D. Gurle, D. Oran. "SIP Extensions for Instant Messaging," IETF. Work in progress.

[draft-rosenberg-sip-3pcc] J. Rosenberg, J. Peterson, H.Schulzrinne, G. Camarillo. "Third Party Call Control in SIP," IETF. Work in progress.

[draft-rosenberg-sip-app-components] J. Rosenberg, P. Mataga, H. Schulzrinne. "An Application Server Component Architecture for SIP," IETF. Work in progress.

[draft-rosenberg-sip-conferencing-models] J. Rosenberg, H. Schulzrinne. "Models for Multi-Party Conferencing in SIP," IETF. Work in progress.

[draft-tsang-appliances-reqs] S. Tsang, S. Moyer, D. Marples, H. Schulzrinne, A. Roychowdhury. "Requirements for Networked Appliances: Wide-Area Access, Control, and Interworking," IETF. Work in progress.

[RFC 768] J. Postel. "User Datagram Protocol," IETF. August 1980.

[RFC 791] J. Postel. "INTERNET PROTOCOL," IETF. September 1981.

[RFC 793] J. Postel. "Transmission Control Protocol," IETF. September 1981.

[RFC 821] J. Postel. "Simple Mail Transfer Protocol," IETF. August 1982.

[RFC 854] J. Postel, J.K. Reynolds. "Telnet Protocol Specification," IETF. May 1983.

[RFC 1633] R. Braden, D. Clark, S. Shenker. "Integrated Services in the Internet Architecture: An Overview," IETF. June 1994.

[RFC 1777] W. Yeong, T. Howes, S. Kille. "Lightweight Directory Access Protocol," IETF. March 1995.

[RFC 1827] R. Atkinson. "IP Encapsulating Security Payload (ESP)," IETF. August 1995.

[RFC 1889] H. Schulzrinne, S. Casner, R. Frederick, V. Jacobson. "RTP: A Transport Protocol for Real-Time Applications," IETF. January 1996.

[RFC 1958] B. Carpenter. "Architectural Principles of the Internet," IETF. June 1996.

[RFC 2026] S. Bradner. "The Internet Standards Process —Revision 3," IETF. October 1996.

[RFC 2045] N. Freed, N. Borenstein. "Multipurpose Internet Mail Extensions (MIME) Part One: Format of Internet Message Bodies," IETF. November 1996.

[RFC 2060] M. Crispin. "Internet Message Access Protocol—Version 4rev1," IETF. December 1996.

[RFC 2068] R. Fielding, J. Gettys, J. Mogul, H. Frystyk, T. Berners-Lee. "Hypertext Transfer Protocol—HTTP/1.1," IETF. January 1997.

[RFC 2131] R. Droms. "Dynamic Host Configuration Protocol," IETF. March 1997.

[RFC 2205] R. Braden, Ed., L. Zhang, S. Berson, S. Herzog, S. Jamin. "Resource ReSerVation Protocol (RSVP)—Version 1 Functional Specification," IETF. September 1997.

[RFC 2246] T. Dierks, C. Allen. "The TLS Protocol Version 1.0," IETF. January 1999.

[RFC 2326] H. Schulzrinne, A. Rao, R. Lanphier. "Real Time Streaming Protocol (RTSP)," IETF. April 1998.

[RFC 2327] M. Handley, V. Jacobson. "SDP: Session Description Protocol," IETF. April 1998.

[RFC 2475] S. Blake, D. Black, M. Carlson, E. Davies, Z. Wang, W. Weiss. "An Architecture for Differentiated Service," IETF. December 1998.

[RFC 2543] M. Handley, H. Schulzrinne, E. Schooler, and J. Rosenberg, "SIP: Session initiation protocol," IETF. March 1999.

[RFC 2597] J. Heinanen, F. Baker, W. Weiss, J. Wroclawski. "Assured Forwarding PHB Group," IETF. June 1999.

[RFC 2598] V. Jacobson, K. Nichols, K. Poduri. "An Expedited Forwarding PHB," IETF. June 1999.

[RFC 2608] E. Guttman, C. Perkins, J. Veizades. "Service Location Protocol, Version 2," IETF. June 1999.

[RFC 2633] B. Ramsdell, "S/MIME Version 3 Message Specification," IETF. June 1999.

[RFC 2705] M. Arango, A. Dugan, I. Elliott, C. Huitema, S. Pickett. "Media Gateway Control Protocol (MGCP) Version 1.0," IETF. October 1999.

[RFC 2779] M. Day, S. Aggarwal, G. Mohr, J. Vincent. "Instant Messaging/Presence Protocol Requirements," IETF. February 2000.

[RFC 2848] S. Petrack, L. Conroy. "The PINT Service Protocol: Extensions to SIP and SDP for IP Access to Telephone Call Services," IETF. June 2000.

[RFC 2960] R. Stewart, Q. Xie, K. Morneault, C. Sharp, H. Schwarzbauer, T. Taylor, I. Rytina, M. Kalla, L. Zhang, V. Paxson. "Stream Control Transmission Protocol," IETF. October 2000.

[RFC 2974] M. Handley, C. Perkins, E. Whelan. "Session Announcement Protocol," IETF. October 2000.

[RFC 2976] S. Donovan, "The SIP INFO Method," IETF. October 2000.

INDEX

E

F

G

H

I

J–L

M

N

O–P

Q–R

S

T

U

V

W–X

ABOUT THE AUTHOR

Gonzalo Camarillo is the Principal Systems Expert with the Advanced Signaling Research Lab of Ericsson in Helsinki, Finland. An active participant in the IETF's SIP Working Group since its inception, he co-authored several contributions on SIP-related matters. He is also a frequent speaker at VoIP conferences and the Ericsson representataive in the SIP Forum. He received an M.Sc. degree in Electrical Engineering from Universidad Politecnica de Madrid and another M.Sc. degree—also in Electrical Engineering—from the Royal Institute of Technology in Stockholm. He is continuing his studies as a Ph.D. candidate at Helsinki University of Technology.